Rural Sector : A View Through Gender Lens

Rural Sector :
A View Through Gender Lens

Dr. Niketha L.
Senior Research Fellow
ICAR-NIANP, Bengaluru-560 030

Dr. Letha Devi G.
Senior Scientist
ICAR-NIANP, Bengaluru-560 030

Dr. P. Adhiguru
Principal Scientist
(Agricultural Extension)
Agricultural Extension Division
KAB-I, Pusa, New Delhi-110 012

SSPH

SATISH SERIAL PUBLISHING HOUSE
403, Express Tower, Commercial Complex
Azadpur, Delhi-110033 (India)
Phone : 011-27672852, Fax : 91-11-27672046
e-mail : info@satishserial.com, hkjain1975@yahoo.com
Website : www.satishserial.com

Published by :

SATISH SERIAL PUBLISHING HOUSE
403, Express Tower, Commercial Complex, Azadpur, Delhi-110033 (INDIA)
Phone : 011-27672852 Fax : 91-11-27672046
E-mail : info@satishserial.com, hkjain1975@yahoo.com

ISBN 978-93-88020-22-0

Composed, Designed & Printed in India

About the Authors

Niketha L., a graduate alumni of the U A S, Bangalore, College of Agriculture, Shimoga (2010), post-graduate in Agriculture Extension from M P K V,Rahuri, College of Agriculture, Pune (2012) and doctoral degree from National Dairy Research Centre, Karnal (2015). Her doctoral research studies bequeathed her with Certificate of Appreciation for 'Outstanding Dissertation Work', besides the supportive fellowship during research tenure and also been accredited with Best Paper Presentation .Her scientific writings had bought her multiple laurels in the field of gender studies, group-cohesion, empowerment, aspirations and farm co-operative organizations, enshrined through myriad national and international journals, popular articles and book chapters.

Dr. Letha Devi G. did her graduation in Agriculture from Kerala Agricultural University in 2001 and topped the university. She completed her Master's in Dairy Extension from National Dairy Research Institute in 2004 and topped the university with Director's Gold Medal. She completed her doctoral degree from Indian Agricultural Research Institute, New Delhi in the discipline of Agricultural Extension in the year 2007. She joined Indian Council of Agricultural Research as a scientist in 2007. She is working as a Scientist under ICAR-National Institute of Animal Nutrition and Physiology, Bangaluru. She did PG Diploma in Intellectual Property Rights from IGNOU, New Delhi in 2013. She has published a number of research papers, and popular articles in various national and international journals of repute. She has published a number of books on various topics related.

Dr. P. Adhiguru Principal Scientist (Agricultural Extension)Agricultural Extension Division, ICAR Head Quarter, Room No. 305, KAB-I, Pusa, New Delhi-110012, India, My significant research management and coordination experience includes dealing with all matters pertaining to Krishi Vigyan Kendras (Agricultural Science Centres), dealing with Research-Extension consequences at different levels, establish linkages with Agricultural Universities, State Governments, NGOs and ICAR Institutes, Inter Ministerial Co-ordination in order to facilitate planning, execution, monitoring and evaluation of the programs being implemented by the Agricultural Extension Division, preparing various technical reports, material, replies for Regional Committees, SFC, General Body / AGM and Parliament Questions and such other Committees, and to assist the Deputy Director General (Agricultural Extension) in all matters relating to the Agricultural Extension.

My significant research project experience includes coordinating 39 ICAR-extramural research projects in extension. I was visiting fellow in Iowa State University, USA and Centre for Biosafety Assessment, Technology and Sustainability (BATS), Switzerland. I was associated with DARE/ICAR for facilitating projects of CABI, IRRI, ICRISAT and other International Collaborative programs with CGIAR Institutes and international Universities, including exchange programs. Involved with various international projects in drafting MoUs and providing inputs related to international agricultural policy. Also, handled projects like Agricultural Science and Technology Indicators (ASTI)" [ICAR-IFPRI], Sustainable rainfed agriculture- impact assessment of World Bank sponsored Watersheds [ICAR-IFPRI-ICRISAT-ODI-CRIDA-World Bank], also, nodal officer for CSISA project of CIMMYT. The other projects include Attracting and Retaining Youth in Agriculture (ARYA), ICT-mediated Agricultural Extension, Geometry of information flow in agriculture, Agricultural-based interventions for sustainable nutritional security, The Upcoming of supermarkets in India: implications for smallholders, Perspectives of Agricultural Extension in India. Visited countries: *Egypt, USA, Switzerland, Indonesia, Malaysia, Philippines, Bhutan, Srilanka and Singapore.*

Preface

Gender is an important consideration in development. It is a way of looking at how social norms and power structures impact on the lives and opportunities available to different groups of men and women. Understanding the gender relations and the power dynamics behind them is a prerequisite for understanding individuals' access to and distribution of resources, the ability to make decisions and the way women and men, boys and girls are affected by political processes and social development.

This book will provide a single platform on the topic related to gender. It covers the concepts and terms related to gender, women role in agriculture, statistical figures, gender issues, gender analysis & its framework, impact assessment, gender mainstreaming, sensitization, gender budgeting in India and also with successful case studies, gender audit, SHGs & its linkage models, digitization of SHGs, women empowerment analytical framework, women development programs, legal frameworks and national policies. In addition, an effort has been made to answer why gender equity is important to achieve gender equality and why gender equality is more important than equity.

Authors

Contents

Chapter 1

Gender in Agriculture

Agriculture is one of the most widespread activities in the world and has a crucial role in food production, environmental protection, landscape preservation, rural employment and food security. Agricultural sector in many developing countries is underperforming, in part because women, who represent a crucial resource in agriculture and the rural economy through their roles as farmers, labourers and entrepreneurs, almost everywhere, face more severe constraints than men in access to productive resources. Agriculture occupies a key position in the Indian economy providing a source of livelihood for a majority of the population. India being a rural and agro-based economy, 54 per cent of its workforce is engaged in agriculture, of which 37 per cent are women. Women provide crucial support in Indian agriculture and household livelihood; however, they remain confined as workers. The national average of land holding by women is 13.5 per cent, varying between 1 and 34 per cent across states (CIWA, 2016).

Agriculture is the mainstay of our economy and women play a significant role in the sustainable development of the economy through their contribution in the household and through agricultural activities, often at par with men. Recognizing the wisdom behind the ancient African proverb "without women we all go hungry," they play a prime role in traditional agriculture, from manually growing crops to agro-processing to homemaking (Majumder J and Shah P, 2017).

Present Status of Women

Women work two-thirds of the working hours, earn only ten per cent of the world's income, own less than one per cent of the world property and two-thirds out of the estimated 876 million adults worldwide who cannot read or write (FAO, 2010). Similarly, the contemporary scenario presents a gloomy depiction in India also. The sex ratio is declining decade by decade where in 1901, it was 972/1000 males but now at present it is 944/1000 male. India has a population of about 1210 million where male accounts for 623.72 (51.54%) and female about 586.47 million (48.46%). About 68.84 (83.3 crore) and 31.16 per cent (37.7 crore) of the population resides in rural and urban areas, respectively. The total literacy rate of India is 74.00 per cent, in which male holds the upper hand with 82.00 per cent literacy and female around 66.00 per cent (Census,GOI, 2011). In India, female headed rural households are about 14.90 per cent only and female agriculture holders are about 10.90 per cent (FAO, 2010). The annual death rate per 1000 population is higher for female children in the age group of 0-4 years. In India, 4.5 million marriages take place every year and out of which, 3 million brides are children resulted in high mortality for girls less than 16 years (Census, 2001). High rate of female foeticide, lower wages, malnutrition, low level of skills, and greater exposure to domestic violence and vulnerability to sexual crimes are some of the other dimensions of women's low status. Out of 128 countries, India ranked 115th in the women empowerment (Anonymous, 2012) which indicates the low status of women in India. The unfair discrimination is more prevalent among rural women due to lack of access to resources and facilities. Though women are not only carriers of human race but also civilisation and sustainable development resets on them. They are the best upholders of environment, ecological and social balances, yet they are victims of male dominated patriarchal system in India. They are neglected in their day-to-day life. Their role is silently appreciated without economic recognition, regard and accountability. They remained as "invisible workers" and have not been given appropriate share in developmental agendas and programmes. They have been largely treated as beneficiaries rather than partners in the developmental process. Further, women are restricted in their access to and control over financial resources and are less involved in financial transactions and decision-making. Less than 45 per cent of women in India have access to funds of their own, and only 15 per cent have a bank or savings account of their own (Kishor and Gupta, 2009). Even though women in rural areas work long hours in and outside the household, e.g., in fuel and fodder collection, fetching water and taking care of animals, their work is often neglected and made invisible in official statistics, as it is not paid-wage labour (Bennett, 1993). So, development and welfare of women have become major concerns of the government as large chunk of population lives in rural India. In this direction, the government has taken various measures to ensure a better status of rural women, to utilise their untapped potential for development and as such to bring them to the mainstream of the development process.

Work Participation of Women

India has approximately 482 million workers, of which 150 million are women; 122 million of them are in the rural areas and the remaining 28 million works in urban areas (Census of India, 2011). Cumulatively, only 7 per cent of India's huge labour force is in the organized sector, which includes workers on regular payrolls, in registered companies and firms (Harriss and Gooptu, 2001). The rest of the workers (93%) work in the unorganised or informal sector. The figures for women workers in India are even more dismal with 96 per cent of them in the unorganized sector. The female work participation rate (WPR) has increased overall from 19.8 per cent in 1981 to 25.5 per cent in 2011. In the rural areas, it has increased from 23.2 to 30 per cent and in the urban areas, it has risen from 8.3 to 15.4 per cent. In agriculture, however, the participation of women has decreased from 71.8 per cent in 2001 to 65 per cent in 2011. As per 2011 Census, 97.6 million females are involved in agriculture, which comprises about 37 per cent of the total agricultural workforce. This is a reduction of around two per cent in comparison to 2001 census. By 2020, women participation is estimated to be around 115 million, that is, 50 per cent in ratio. Though female literacy rate has increased from 8.9 per cent in 1951 to 65.5 per cent in 2011 growth is sluggish in comparison to their male counterparts (Majumder J and Shah P, 2017).

Women in Agriculture

The global average is dominated by Asia. Within Asia, the sub-regional averages range from about 35 per cent in South Asia to almost 50 per cent in East and South-east Asia. The Asian average is dominated by China, where the female share of the agricultural labour force has increased slightly during the past three decades. The female share in India has remained steady at just over 30 per cent. Female share of the agricultural labour force appears to have increased dramatically, now exceeding 50 per cent in Bangladesh. Other Asian countries, such as Malaysia have seen declining female labour shares in agriculture (FAO, 2011).

Agriculture is much more important for women than for men in terms of employment in South Asia and the Middle East. It is somewhat more important for women than for men in East Asia, North Africa and sub-Saharan Africa. In Central and South Eastern Europe and in Latin America, women are much more concentrated in the service sector.

In India, agriculture is a highly gender sensitive sector. 75.38 percent of all women workforce are engaged in agriculture. Within agriculture, 94 percent of women in crop cultivation are in cereal production and other crops n.e.c, 1.4 percent in vegetable production and horticulture, while 3.72 percent are engaged in fruits, nuts, beverages, and spice crops. Women's participation rate (% share in all workers) is high in tea plantations (47.08%), cotton cultivation (46.84%) and growing oilseeds (45.43%). Vegetables also show a high share of women workers (39.13%). All these crops need labour intensive work. Foodgrains production too draws about 33 per cent of its labour from women while growing of sugarcane and sugar beet draws 25.5 per cent. Women are also significantly engaged in agricultural allied activities, for example, in animal husbandry (7.03% of women engaged in agriculture & allied activities, and 71.08% of women engaged in agricultural allied activities), plantations and fisheries. These are also significant livelihood options for women in specific regions in India. Indian agriculture, with its low requirement of skills, and work which can be more easily combined with work at home, is an easy source of work for women, though women are confined to low end jobs like sowing, transplanting, weeding and harvesting. Many women also work as unpaid family labour. While gender wage disparity in all these activities is about the same, with women earning only 70 per cent of men's wages, the gap between wages for ploughing earned by males and weeding wages earned by females is Rs. 21.47. The level of lower skill and education level is highlighted by the fact that 52-75 per cent of women engaged in agriculture are illiterate. This also implies that they are less able to shift easily to other higher skilled jobs, for example, in the services sector. This makes them dependent on agriculture and on its stable growth for survival. The distribution of critical resources like land is also unevenly distributed across gender. Women seldom enjoy property ownership rights directly in their names. Even when women have mutations of land in their names, they may not have actual control over that land. Decision-making in cropping patterns, sale, mortgage and the purchase of land or the instruments of production remains in the hands of the men of the household. With unequal access to other resources like credit, skills, and other inputs aggravating this deep inequality, land inequality is one area that is of crucial importance for rural women in India and goes a long way towards strengthening the unequal socio-economic relations in our society.

More than half of world's food is grown by women, but worldwide her hard work has mostly been unpaid (Dankelman and Davidson, 2013). Women's indigenous knowledge and skills are vitally necessary for food production and sustainable agriculture (Singh and Arora, 2010). The need to accept the women's role is not merely a matter of justice. The constraint imposed upon women by gender bias has aggravated an already truncated production. Burden on women is getting doubled, as more and more men are moving into other remunerative sectors, leaving agriculture in the hands of women. In a study, two-thirds of the women felt that they had to work harder to receive the benefits as comparably employed man. Women are not perceived as "farmers" even when they do most of the farm work. This is supported by the fact that there are 5 per cent more female marginal agricultural labourers than the male labourers (Census of India, 2011). Interestingly, agricultural extension and information on new technologies are almost exclusively directed to men, even though women are increasingly responsible for farm work (Kelkar, 2011). Two decades back, researchers revealed that the increasing use of new technological devices resulted in an inevitable decline in employment opportunities for women (Mencher and Amico, 1986). Women due to their passive role, lack of knowledge, ignorance, accessibility, and outreach, as well as resistance from men in the society and family have consistent inaccessibility to modern technology of cultivation and other facilities available to their male counterparts. Adding to the irony is the incompetence of women farmers in conversation, banking systems, and procedures for accessing loans and thus, remains deprived of credit facilities for the purchase of machinery and other agricultural input. The women workers seldom handle the purchase or sale of agro-produce within their village boundaries through trade in village market, but they require knowledge of industries dealing in machinery and organized marketing network system for better and profitable marketing of agro-produce. However, the socio-cultural bindings at home and work limit their full-fledged participation and recognition (Majumder J and Shah P, 2017).

Women Involvement in Livestock

Livestock are important in women's livelihoods and assets portfolios. Two-third of the world 600 million poor livestock keepers are rural women (Thornton *et al.*, 2003). Dairying is an occupation that supports the livelihood of women, especially the rural poor women in India. Almost 70 per cent of the dairy farmers are landless, small or marginal farmers, who own only one or two cattle. Landless farmers dependent on livestock to some extent or completely form a relatively large proportion in India (Orskov, 1993). Women are traditionally responsible for milking of animals, processing of milk and collecting dairy products (FAO, 2010). Seventy million rural households were engaged in milk production, the majority being small and marginal farmers and landless. Dairy co-operatives ensure inclusiveness and livelihoods for small holders, especially women (Anonymous, 2012).

There is a great variability across systems and socio-economic context; women generally play a major role in managing and caring for animals, even when they are not owners. As clearly pointed out by Flintan (2008) that woman participated in every aspects of livestock management in different pastoral systems around the world. In intensive Asian livestock systems, more than three-quarters of livestock-related tasks are the responsibility of women (Niamer, 1994). Indian women play a significant role by keeping livestock by providing labour, in poorer families; their contribution typically exceeds that of men (George and Chokshi, 1990). Further, dairy farming is considered as an extension of domestic activities, like–feeding and watering of animals, bringing of fodder from the field, cleaning of animals and sheds, preparation of cow-dung cakes, protection of animal from diseases, milking, milk products making and marketing of milk. Dairying in India is a female dominated enterprise (Fulzele and Meena, 1995). It is established beyond doubt that women always participated in dairy and animal husbandry activities in addition to their daily household chores (Belurkar *et al.*, 2003). Women play crucial and significant role in livestock rearing, but their contribution in livestock rearing has not been given the due place they deserve. They always remain invisible workers (Chayal *et al.*, 2009). Moreover, in India, majority of dairy farm women carried out activities, like care of newborn calf, milking, cleaning of animal shed, cleaning of utensils, weaning and management of calf, preparation of cow-dung cakes and construction of animal sheds but their participation was least in maintenance of farm records. Involvement of farm women in the care of newborn calf and cleaning of utensils and shed was 100 per cent (Sah *et al.,* 2006; Chayal *et al.,* 2009 and Lahoti *et al.,* 2012). Harvest and transport of grass for livestock was performed exclusively by women (Johnson *et al.,* 2013). The majority (73.33%) of farm women were involved in compost making followed by 70.00 per cent and 66.66 per cent of them carried out milking of animals and weaning & management of calves, respectively (Lahoti *et al.,* 2012). Rathod *et al.,* (2011) also reported that women performed activities, like milking (90%), cleaning of animal sheds (89.16%) and disposal of cow-dung or preparation of cow-dung cakes (86.66%). About 75.83 per cent of the farm women were involved in the construction of animal sheds (Toppo *et al.,* 2004 and Rathod *et al.,* 2011). Most of the activities of dairy farm were carried out by women, so they spent more time compared to men in dairy production. Farm women spent about 294.34 min. daily in different dairy farm activities (John and Thirunavukkarasu, 2002).

Gender is embedded in the distribution of essentially all the resources used in agriculture including land ownership, farm management decisions, market access for inputs and outputs, information from extension services, use of information and communication technology, etc. Gender is also embedded in the distribution of the gains from increased agricultural productivity, particularly influencing who controls the outputs and decides how the proceeds will be used. If interventions fail to consider how gender is embedded in the system, we will miss critical opportunities for transforming agricultural systems, increasing productivity, reducing poverty and improving people's lives.

❑❑❑

Chapter 2

Gender Concepts and Terms

Gender

The word 'gender' is used to describe the characteristics, roles and responsibilities of women and men, boys and girls, which are socially constructed. Gender is related to how we are perceived and expected to think and act as women and men because of the way society is organized, not because of our biological differences.

Definition of Gender

Gender is defined by FAO as 'the relations between men and women, both perceptual and material. Gender is not determined biologically, as a result of sexual characteristics of either women or men, but is constructed socially. It is a central organizing principle of societies, and often governs the processes of production and reproduction, consumption and distribution' (FAO, 1997).

Gender has also been defined as: "The commonly shared expectations and norms within a society about appropriate male and female behaviour, characteristics and roles. Gender can be considered a social and cultural construct that differentiates females from males and thus, defines the ways in which females and males interact with each other. These roles and expectations are learned and they can change over time as well as vary within and between cultures."

Difference between Gender and Sex

The terms 'sex' and 'gender' are closely linked, yet they are not synonyms. Robert Stoller, in the 1960s, has drawn the distinction between them. He suggested that the word 'sex'' be used to refer to the physical differences between men and women, while the term 'gender' be used in connection to the behaviour and cultural practices of men and women.

Sex: 'Sex' refers to the biological characteristics or natural biological differences between men and women.

Gender: Gender refers to the cultural, socially constructed differences between the two sexes. It refers to the way society encourages and teaches the two sexes to behave in different through socialization. 'Gender' and the hierarchical power relations between women and men based on this are socially constructed, and not derived directly from biology. Gender identities and associated expectations of roles and responsibilities are therefore, changeable between and within cultures.

Gender Dynamics: Gender dynamics include the relationships and interactions between and among boys, girls, women and men. Gender dynamics are informed by socio-cultural ideas about gender and the power relationships that define them. Depending upon how they are manifested, gender dynamics can reinforce or challenge existing norms. Gender dynamics is nothing but the way in which men and women are treated or behave differently in society, either with their own gender or with each other. The changes in gender dynamics and roles in society for an individual or group can be changed by either economic standing, age or other factors.

Gender narratives can be identified in the development discourse: the Women in Development (WID) approach, the Gender and Development (GAD) approach.

Women in Development Approach: In the late 1970s and early 1980s, the WID approach became popular in the development field as a result of the concern that women were being left out of economic development processes. The approach focused on the inclusion of women in development as a tool to increase the economic and social efficiency of development processes. The WID approach is prominently featured in the concept of gender parity for example, the notion that an equal proportion of girls and boys should be enrolled in and complete schooling.

Gender and Development Approach: By the late 1980s, the GAD approach came to the forefront. This approach sought to challenge root causes of gender inequality and increase women's access to resources and decision-making (DFID, 2002). The GAD approach focused on 'empowerment', or increasing the agency of women and giving them the power to take control of their own lives (UNESCO, 2009), often through targeted trainings and workshops. This approach also emphasizes gender equity as an objective, which refers to the process of being fair to women & men and challenging policies that unfairly bias men or women.

Shift in development approach: WID vs. GAD

Criteria	Women in Development (WID)	Gender and Development (GAD)
The Approach	An approach which views women's lack of participation as the problem	An approach to people centered development
Focus	Women	Relations between women and men
The Problem	The exclusion of women (half of the productive resource) from the development process	Unequal relations (between women and men, rich and poor) that prevents equitable development and women's full participation
The Goal	More efficient, effective development	Equitable, sustainable development with men and women sharing decision-making and power
The Solution	Integrate women into existing structures	Empower the disadvantaged and women Transform unequal relations and structures
The Strategies	Women only projects • Women's component integrated projects • Increase women's productivity • Increase women's income • Increase women's ability to manage the household	Identify/address practical needs determined by women and men to improve their conditions • At the same time address strategic gender needs of women and men • Address strategic needs of the poor through people centered development

Gender Identity

It is the sense of people as male or female. It is related to our existence as male or female. The sense of identity creates interest in their life. For instance, children imbibe the values and norms of the elderly people. It affects their sense of gender identity. Social construction of gender as well as the sexual difference are very much part of the process of gender identity. It is recognition of one's own identity. It is conditioned according to the dominant norms and values of the society.

Gender Discrimination

Prejudicial treatment of an individual based on a gender stereotype (often referred to as sexism or sexual discrimination).

Gender Management System (GMS)

Gender management system is a network of structures, mechanisms and processes put in place within an existing organizational work to guide, plan, monitor and evaluate the process of mainstreaming gender into all areas of the organizational work, in order to achieve greater equality and equity within the context of sustainable development. A GMS may be established at any level of Government or in institutions, such as Universities, Inter-governmental or Non-Governmental Organizations, private sector organizations are trade unions.

Gender Mainstreaming

The concept of gender mainstreaming was first proposed at the 1985 Third World Conference on Women in Nairobi. The idea has been developed in the United Nations development community. The idea was formally featured in 1995 at the Fourth World Conference on Women in Beijing.

United Nations has defined Gender Mainstreaming as the process of assessing the implications for women and men of any planned action, including legislation, policies or programmes, in all areas and at all levels. It is a strategy for making women's as well as men's concerns and experiences an integral dimension of the design, implementation, monitoring and evaluation of policies and programmes in all political, economic and societal spheres so that women and men benefit equally and inequality is not perpetuated.

Gender mainstreaming is not only a question of social justice, but is necessary for ensuring equitable sustainable human development by the most effective and efficient means.

Gender Planning

Gender planning is the recognition of existing gender inequalities in the society and helps to formulate the policies to mitigate gender inequalities in the society. It aims to improve the status and conditions of women by formulating appropriate policies and programmes.

Gender Budgeting

The council of Europe defines gender budgeting as a gender based assessment of budgets incorporating a gender perspective at all levels of the budgetary process and restructuring revenues and expenditures in order to promote gender equality.

Gender Responsive Budgeting

A Gender Responsive Budget is a budget that acknowledges the gender patterns in the society and allocates money to implement policies and programmes that changes these patterns in a way that moves towards a more gender-equal society.

Gender Auditing

Gender auditing is the analysis of the system and process of finance of any government. Gender auditing implies auditing the income and expenditure of governments from a gender perspective and also analyzing the development process including the process of legislation, guidelines, taxes and social development projects.

Gender-Blindness

Gender-blindness refers to a failure to identify or acknowledge difference on the basis of gender where it is significant. This approach does not recognize gender as an essential factor that determines the choices for men and women separately in the society.

Gender Policy

A gender policy is an integral part of all its organizational policies, programmes and projects and involves building a culture that understands the issues and policies, which respect diversity and gender, related concerns. The policy seeks to ensure greater consistency of gender principles, policies and practices across the organisation and to provide an accountability framework in relation to gender, against which all staff can be accountable and against which the organization can audit itself.

Gender Roles

The term 'gender role' was first coined by John Money in 1955. The term 'gender role' refers to society's concept of how men and women are expected to act and how they should behave. These roles are based on norms, or standards, created by society.

Gender Oppression

Where one gender dominates the other unjustly or even cruelly, whether this happens deliberately or not.

Types of Gender Roles

Gender roles are a specific set of social and behavioural actions which are considered to be appropriate for the given gender. The concept of gender roles has been developed from the work of Caroline Moser. She explains this concept as follows:

Gender planning recognises that in most societies low-income women have a triple role: women undertake reproductive, productive, and community managing activities, while men primarily undertake productive and community politics activities.

Reproductive Role	Childbearing/rearing responsibilities and domestic tasks done by women, required to guarantee the maintenance and reproduction of the labour force. It includes not only biological reproduction but also the care and maintenance of the workforce (male partner and working children) and the future workforce (infants and school-going children).
Productive Role	Work done by both men and women for pay in cash or kind. It includes both market production with an exchange-value, and subsistence/home production with actual use-value, and also

	potential exchange-value. For women in agricultural production, this includes work as independent farmers, peasant wives and wage workers.
Community Managing Role	Activities undertaken primarily by women at the community level, as an extension of their reproductive role, to ensure the provision and maintenance of scarce resources of collective consumption, such as water, health care and education. This is voluntary unpaid work, undertaken in 'free' time.
Community Politics Role	Activities undertaken primarily by men at the community level, organising at the formal political level, often within the framework of national politics. This is usually paid work, either directly or indirectly, through status or power.
Multiple Roles	Both men and women play multiple roles. The major difference, however, is that • Men: typically play their roles sequentially, focusing on a single productive role • Women: must usually play their roles simultaneously, balancing the demands of each within their limited time constraints.

Gender Needs

Gender needs arise because the needs of men and women are different, i.e. men and women have different gender roles, they do different types of work, have different degrees of access to services and resources, and they experience unequal relations. In gender analysis a distinction is made between practical gender needs and strategic gender needs.

i) **Practical gender needs**

- Related to peoples basic needs for their day-to-day survival.
- Concerned with improving the condition of women and men through promoting the efficiency of resource use, rather than addressing issue of empowerment.
- In development programmes these needs may include the provision of services, such as clean water, shelter and health

care as well as income generating opportunities. Meeting practical needs means giving response to immediate perceived necessities.

ii) **Strategic gender needs**

- Refers to the relative status/position of women/men in pursuit of empowerment and equity.
- Strategic interests may include legal rights, increased decision making, protection from domestic violence, etc.
- Practical and strategic interests are complementary (projects that only target practical needs may not be sustainable unless strategic interests are also taken into account).

Practical needs	**Strategic needs**
Tend to be immediate & short-term	End to be long-term
Related to daily needs: food, housing, income, health, etc.	Related to a subordinate position: lack of resources and of education and training, vulnerability to poverty and violence, etc.
Can be satisfied by specific inputs (provision of food, technologies, health facilities, etc.)	Can be satisfied by confidence building, improved self-confidence, education, political mobilization, strengthening of women's organisations, etc.
Can improve women's living conditions	Can improve the situation of women has in the society.
Generally, does not change the traditional roles and social relations.	Can give more power to women and transform the social relations.

Quick Definitions

Culture	The distinctive patterns of ideas, beliefs, and norms which characterise the way of life and relations of a society or group within a society.
Sex and Gender	Sex refers to the biological characteristics that categorise someone as either female or male; whereas gender refers to the socially determined ideas and practices of what it is to be female or male.

Gender Roles	The particular economic, political and social roles and responsibilities that are considered appropriate for men and women in a culture.
Gender Equality	The absence of discrimination on the basis of a person's sex in authority, opportunities, allocation of resources or benefits, and access to services.
Gender Equity	The process of being fair to women and men. Equity leads to equality.
Gender Parity	A numerical concept concerned with the relative equality in terms of numbers and proportions of women, men, girls and boys. In education, this means that the same number of boys and girls receive educational services at different levels and in diverse forms.
Gender Blindness	Failure to recognise that the roles and responsibilities of women/girls and men/boys are ascribed to, or imposed upon them in specific social, cultural, economic and political contexts.
Gender Analysis	The systematic gathering and examination of information on gender differences and social relations in order to identify, understand and redress inequities based on gender.
Gender Discrimin-ation	The systematic, unfavourable treatment of individuals on the basis of their gender, which denies them their rights, opportunities or resources.
Gender Division of Labour	The socially determined ideas and practices which define what roles and activities are deemed appropriate for women and men.
Gender Equality	Gender equality denotes women having the same opportunities in life as men, including the ability to participate in the public sphere.
Gender Equity	Gender equity denotes the equivalence in life outcomes for women and men, recognising their different needs and interests, and requiring a redistribution of power and resources.
Gender Awareness	Understanding that there are socially determined differences between men and women, and that these influence access to and control of resources.
Gender Needs	Shared and prioritised needs identified by women that arise from their common experiences as a gender.

Gender Planning The technical and political processes and procedures necessary to implement gender-sensitive policy.

Gender Training A facilitated process of developing awareness and capacity on gender issues, to bring about personal or organisational change for gender equality.

Gender Budgetting Application of gender mainstreaming in the budgetary process. It entails a gender based assessment of budgets, incorporating a gender perspective at all levels of the budgetary process & restructuring revenues and expenditures in order to promote gender equality.

Gender Analysis It is a systematic methodology for examining the differences in roles & norms for women & men, girls & boys, the different levels of power they hold, their differing needs, constaints and opportunities, and the impact of these differences in their lives.

WID/GAD The WID (Women in Development) approach calls for greater attention to women in development policy and practice, and emphasises the need to integrate them into the development process.

In contrast, the GAD (Gender and Development) approach focuses on the socially constructed basis of differences between men and women and emphasises the need to challenge existing gender roles and relations.

Gender Audit Assessment of the extent to which gender equality is effectively institutionalised in policies, programmes, organisational structures and proceedings (including decision-making processes), and in the corresponding budgets.

Gender Awareness Raising Process that aims at showing how existing values and norms influence our picture of reality, perpetuate stereotypes and support mechanisms (re)producing inequality.

Gender Balance Human resources and equal participation of women and men in all areas of work, projects or programmes.

Gender Contract Set of implicit and explicit rules governing gender relations, and which allocate different work, value, responsibilities and obligations to women and men, and are maintained on three levels: cultural superstructure (the norms and values of society); institutions (family welfare, education and employment systems, etc.); and socialisation processes, notably in the family.

Gender Relations	Specific subset of social relations uniting women and men as social groups in a particular community, including how power - and access to/control over resources - is distributed between the sexes.
Gender Dimension	Ways in which the situation and needs of, and challenges facing, women and men (and girls and boys) differ, with a view to eliminate inequalities and avoiding their perpetuation, as well as to promote gender equality within a particular policy, programme or procedure.
Gender Disparities	Differences in women's and men's access to resources, status and well-being, which usually favour men and are often institutionalised through law, justice and social norms.
Gender Dynamics	Relationships and interactions between and among girls, boys, women and men. Gender dynamics are informed by socio-cultural ideas about gender & the power relationships that define them. Depending upon how they are manifested, gender dynamics can reinforce or challenge existing norms.
Gender Education	Necessary part of curricula at all levels of the education system, which would enable both girls and boys, women and men to understand how constructions of masculinities and femininities and models for assigning social roles - which shape our societies - influence their lives, relationships, life choices, career trajectories, etc.
Gender Equality Compe-tence	Skills, attributes and behaviours that people need in order to mainstream gender concerns effectively into policies it plans and helps to build gender equality.
Gender Equality Competence Develop-ment	Series of activities that aim at strengthening people's skills & knowledge on a certain matter in this case gender equality.
Gender Equality Training	Any educational tool or process that aims to make policy-makers and other actors more aware of gender equality issues, build their gender competence and enable them to promote gender equality goals in their work at all levels.
Gender Stereotypes	Preconceived ideas whereby females and males are arbitarily assigned characteristics and roles determined and limited by their gender.

Gender Evaluation Evaluation applied as a method of gender mainstreaming that integrates gender equality concerns into the evaluation objectives, evaluation methodology, approaches and use.

Gender Inequality Legal, social and cultural situation in which sex and/or gender determine different rights and dignity for women and men, which are reflected in their unequal access to or enjoyment of rights, as well as the assumption of stereotyped social and cultural roles.

Gender Expression People's manifestation of their gender identity, and the one that is perceived by others.

Gender Gap Gap in any area between women and men in terms of their levels of participation, access, rights, remuneration or benefits.

Gender Identity Each person's deeply felt internal and individual experience of gender, which may or may not correspond to the sex assigned at birth, including the personal sense of the body (which may involve, if freely chosen, modification of bodily appearance or function by medical, surgical or other means) and other expressions of gender, including dress, speech and mannerisms.

Gender Norms Standards and expectations to which women and men generally conform, within a range that defines a particular society, culture and community at that point in time.

Gender Parity Numerical concept related to gender equality. Gender parity concerns relative equality in terms of numbers and proportions of women and men, girls and boys, and is often calculated as the ratio of female-to-male values for a given indicator. When males-to-females ratios are calculated instead, the label 'sex ratio' is used instead of 'gender parity'.In the context of gender equality, gender parity refers to the equal contribution of women and men to every dimension of life, whether private or public.

Gender Perspective Taking into account gender-based differences when looking at any social phenomenon, policy or process. The gender perspective focuses particularly on gender-based differences in status and power, and considers how such discrimination shapes the immediate needs, as well as the long-term interests, of women and men.

Gender-Neutral Policy, programme or situation that has no differential positive or negative impact in terms of gender relations or equality between women and men.

Gender-Sensitive Policies and programmes that take into account the particularities pertaining to the lives of both women and men, while aiming to eliminate inequalities and promote gender equality, including an equal distribution of resources, therefore addressing and taking into account the gender dimension.

Gender Sensitivity	Aim of understanding and taking account of the societal and cultural factors involved in gender based exclusion & discrimination in the most diverse spheres of public and private life.
Women's Empower -ment	A 'bottom-up' process of transforming gender power relations, through individuals or groups developing awareness of women's subordination and building their capacity to challenge it.
Gender bias	Gender bias is behaviour that shows favouritism towards one gender over another. Most often, gender bias is the act of favouring men and/or boys over women and/or girls.

❑❑❑

Chapter 3

Gender Issues

The presence of dynamic integraction between gender and development, create a growing concern by national and international governments and communities. We live in a gendered world. Everywhere we look, we see differences between how males and females are situated in society, what they do, where they gather in groups, to what degree they have access to and control over resources, etc. Although it is easy to accept as fact that men and women are actually different in some ways and to appreciate the differences, what is increasingly diffcult to accept is that women around the world are discriminated against in particular way simply because they were born female.

Agricultural extension programmes ensure that information on new technologies, plant varieties and cultural practices reaches farmers. However, in the developing world it is a common practice to direct extension and training services primarily towards men. A recent FAO survey showed that female farmers receive only five per cent of all agricultural extension services worldwide and that only 15 per cent of the world's extension agents are women.

Studies on agricultural extension have highlighted a number of weaknesses in reaching rural women. Traditionally, most extension services have been devoted to farmers who own land and who are willing and able to obtain credit and invest it in inputs and technological innovations. Since women often lack access to land or access to other collateral with which to obtain credit, extension services, unintentionally, bypass women.

For too long, policies have been based on the assumption - proved wrong by studies - that information conveyed to the male head of a household would be passed on to its female members. But men do not necessarily discuss production decisions with their wives or transfer extension knowledge to them. Furthermore, policy-makers fail to recognize that men and women are often responsible for different crops, livestock, tasks and income-generating activities and that their extension needs consequently differ.

Extension services usually focus on commercial production rather than on subsistence crops, which are the primary concern of women farmers and also the key to food security in developing countries. Extension agents will often choose to work with a few farmers judged to have a progressive attitude, while neglecting the resource-poor and landless, including women. To compound the problem, extension meetings are often scheduled at times when women farmers are unable to attend because of their other household responsibilities.

As rural women are a vital link in agricultural development, it is essential that they take their place alongside men as full participants in and beneficiaries of extension programmes.

Gender Issues in Agriculture

Over-burden of Work

Rural women are much more over-burdened than men owing to their multiple occupations. Researches on women in agriculture have revealed that on an average women work for 15-16 hours a day. Studies further point out that farm activities, which are time and labour intensive, monotonous, repetitive and more drudgery prone, are generally performed by women. Since these operations are done manually, they cause considerable physical and mental fatigue and health problems.

Impact of Technology

Some of the new agricultural technologies are reported to have affected farm women adversely. Green revolution had led to the dispossession of small women land-holders, who have been forced to join the ranks of wage earners. Wherever the new agricultural technology led to multiple cropping, the work load of women has increased. While a number of tasks performed by males have been mechanized, the tasks usually allotted to women continue to be manual and suffer from drudgery. Even where improved techniques have been found for the women's activities, there is not sufficient access to training in such techniques.

Facilities and Support Services

There is rigidity for female labourers in terms of working hours, place and duration of work. Because of this, children are neglected and health of women is also adversely affected. Lack of adequate support services like child care services-creche, balwadi, adequate maternity and health care- lack of safe drinking water, etc. further add to their problems.

Development Bias

Despite the contribution of women in the production process, persistent bias of development planners in treating them primarily as consumers of social services rather than producers, kept them away from the development programmes in agriculture and allied sectors.

Women suffer from a statistical *purdah* as a result of which their contribution is not recognized. They often have heavier workloads than men and bear virtually sole responsibility for family welfare and household management. However, they have limited control over productive resources. Gender discrimination, rooted in law and custom, is pervasive and impedes socio-economic development.

Constraints to Women's Access to Resources

Many of the constraints that rural women confront are similar to those all resource poor farmers confront, such as lack of access to land, credit, training, extension and marketing facilities. But, for social and economic reasons, women's constraints are even more pronounced and, in general, development interventions that seek to remove constraints for poor farmers do not reach women.

Consequently, the development of technologies specifically tailored to women-specific occupations and the involvement of women in technology development and transfer have received inadequate attention from both scientific and administrative departments of governments.

Access to Land

Women's lack of access to land or insecure tenure continues to be a major obstacle to increase their contributions and benefits. When women have access to land, they often do not have secure tenancy and tend to have smaller and less productive plots in comparison with men. While land access is increasingly problematic for poor men and women, women's access has further restriction by inheritance laws and customs. Rural women-headed households are especially affected by land constraints.

Access to Credit

Women are better credit risks. In places where women are legally entitled to access financial institutions, they face problems getting loans because they often belong to the poorest sector of the rural population. Rural financial institutions are also often hesitant to accept women clients because they, in an even greater proportion than men, cannot fulfill collateral requirements, are inexperienced borrowers do not have access to extension and marketing services and cannot fulfill application requirements.

Access to Markets

Women engaged in agriculture, forestry and fishery tend to produce small quantities and have poor access to marketing boards and co-operatives. Therefore, women sell mainly to private traders and have low bargaining power. Institutions which promote women's group access to market should be strengthened. Successful examples are of SEWA, Gramin Bank, SHG Federations, etc.

Research and Technology Development

Women only benefit from agricultural support programmes if the information, technology and methods imparted are relevant to their production activities. Agricultural research is generally very less oriented towards adapting technology to women's physical capabilities or towards addressing their tasks. Women's low productivity stems mainly from lack of appropriate technology.

Access to Extension and Training

Women farmers usually have been neglected in extension efforts. Recently, the need for innovating changes in extension programmes for women farmers is being felt. The Central Sector Scheme, "Women in Agriculture" which started on a pilot basis in 1993- 94 in seven states, has shown encouraging results and was expanded to a few other states. The need for capacity building and skill up-gradation of farm women is now receiving the priority it deserves. Special extension and technology dissemination programmes for women are being implemented.

Specific consequences of the inequality between women and men

1. Traditionally women are expected to perform work which involves manual, respective tasks.
2. Men posses and control land
3. Women's work is underpaid and undervalued
4. Women's work is unaccounted
5. Women undertake multiple workload at home, farm and community.
6. Women lack access to skills, trainings and information
7. Women get displaced from traditional employment opportunities.
8. When technology is introduced and men take over the task
9. Women lacks access to firm inputs, credit and financial incentives
10. Women lacks access to market and control over income
11. Women face food insecurity
12. Women's priorities, needs, problems are unattended and neglected
13. Farmer is regarded as male
14. Women are discriminated in design and testing of technologies.

Challenges for development of Gender

1. Gender stereotypes
2. Gender bias
3. Gender discrimination
4. Gender devision of labour

To overcome from this challenges we need to take up the following methods:

1. Gender sensitization
2. Gender mainstreaming

3. Gender budgeting
4. Gender equity
5. Gender equality

Advantages from these methods

1. Looking at gender differences by identifying tasks, activities and rewards associated with the gender division of labour
2. Bridging agricultural technological gap between women and men farmers
3. Bridging agricultural information gap
4. Focusing on agricultural interventions from gender perceptive
5. Examining budget from a gender perceptive
6. Looking into programmes/project activities from a gender lens
7. Developing action plans from a gender perceptive
8. Mainstreaming gender into developmental activities including training, transfer of technology mechanism and other extension services.

❏❏❏

Chapter 4

Status of Women in Economy and Barriers to Gender Equality

Women in the Economy

The participation of women in the workforce, the quality of work allotted to them and their contribution to the GDP are indicators of the extent of their being mainstreamed into the economy. On all these parameters women in India are worse than men and the challenge is to bridge the inequality. Opening up of the economy and rapid economic growth have escalated some of the existing structural barriers faced by women and new challenges in the form of dismantling of traditional support structures, displacement due to migration, obsolescence of traditional skill sets have emerged.

Data from the 66th round of the NSSO indicates that female work participation rate has decreased between 2004-05 to 2009-10. The share of women in usual status workers declined from 28.7% to 22.8%. In rural areas this has declined from 32.7% to 26.1% and in urban areas from 16.6% to 13.8%.

Participation in the Labour Force

Indicators	Rural (%)			Urban (%)			Total
	Male	Female	Total	Male	Female	Total	Persons
Labour Force Participation Rate	55.6	26.5	41.4	55.9	14.6	36.2	40.0
Work Participation Rate	54.7	26.1	40.8	54.3	13.8	35.0	39.2
Proportion Unemployed	0.9	0.4	0.7	1.6	0.8	1.2	0.8
Unemployment Rate	1.6	1.6	1.6	2.8	5.7	3.4	2.0

Source: NSSO (2009-10)

A little under four-fifths (about 79 per cent) of the rural women workers are employed in the agriculture sector , a sector that is marked by shortage of paid jobs, decelerating and differential wages on basis of gender and degradation of resources. It is estimated that about 60% of all agricultural operations are handled exclusively by women. Female hourly wage rates in agriculture vary from 50 to 75% of male rates, and are too low to overcome absolute poverty.

Situation Analysis and Barriers to Gender Equality

Empowerment of women is closely linked to the opportunities they have in education, health, employment and for political participation. Over the years, significant advancements have been made in India on many of these counts. Data on literacy rates, enrolment and drop rates in primary education, life expectancy, infant mortality, maternal mortality rates, etc has shown a progressive trend. However, other parameter that reflects the status and position of women in society, such as work participation rates, sex ratio in the age group of 0-6 years and gender based violence continue to be heavily skewed against women.

Selected Indicators on Status of Women

Indicators	Male	Female
Literacy Rate (%) Census, 2011	82.14	65.46
Sex Ratio Census 2011	1000	940
Child Sex Ratio (0-6 years) Census, 2011	1000	914
Worker Population ratios (Per 1000)	819	336
MPs in Lok Sabha (%)	89.18	10.82

Violence against Women

It is widely acknowledged that deep-rooted ideologies of gender bias and discrimination - the confinement of women to the private domestic realm, restrictions on their mobility and exclusion from the public political sphere continue to daunt the majority; and the entitlements and public services, which constitute the poor women's life line, do not reach them. Such social and structural barriers to women's empowerment manifest themselves in various ways.

There is a violence against women- in the home and outside. Violence against a woman affects her sense of self-esteem, demolishes her self-confidence and is often used as a potent tool of subjugation and disempowerment. The National Family Health Survey (NFHS-

III, 2005-06) reported that *one-third* of women *aged 15 to 49* had experienced *physical violence*, and approximately *one in 10* had been a victim of *sexual violence*. The survey also found that only one in four abused women had ever sought help, and that 54% of women believed it was justified for a husband to beat his wife.

A study of the data from National Crime Records Bureau (NCRB) shows the increasing incidence of crimes against women. The total number of crimes against women increased by 29.6 per cent between 2006 and 2010. Further, these numbers have to be viewed keeping in mind that not all crimes against women are reported. The actual numbers may give even greater cause for concern. What is more disturbing is that conviction rates remained low, reflecting, inter alia, that many of these cases are not being prosecuted properly and inadequate proof is tendered before the courts.

Early marriage is another area of concern which makes women more vulnerable to domestic violence. According to the NFHS III data, the median age of marriage for women in the 20-49 years age group, ranges between 16.5 years to 18.3 years. The data also shows that more than one-quarter (27%) of the women in this age group got married before their 15th birthday and more than half (58%) got married before the legal minimum marriage age of 18. The data also reflects wide disparities across States, with the proportion marrying that early being as high as 60-61% in the States of Jharkhand and Bihar, and as low as 12% in Himachal Pradesh and Goa - the States with the lowest and highest educational levels, respectively. This suggests an inverse relationship between education and early marriage. There is evidence that marriage before the age of 18 constraints adolescent girls' opportunities to obtain higher education, and severely restricts their autonomy. Further, the decline of this "early marriage rate" has been slow, of only 5 per cent points from 50 per cent of all women in 1993 to 45 per cent in 2006. Marital abuse is also a common problem in such marriages.

Another consequence of gender discrimination, especially violence against women, in India is the declining child sex ratio. This phenomenon has spread across States. The census data has revealed that the worst child sex ratios have been found in States with high per capita incomes, while, States with low per capita incomes like Bihar, Rajasthan and UP continue to suffer from adverse sex ratios due to persisting gender discrimination. The declining child sex ratio is a silent demographic disaster in the making which will have adverse implications on women in the form of increased trafficking for sexual exploitation, honour killings, 'bought' brides, rapes, etc. Already, States such as Haryana and Punjab which have acutely adverse child sex ratios are displaying disturbing trends.

Indicators like the adverse sex ratio (940 / 1000) and the less than five mortality rate for girls in India (73 per 1000 live births as compared to 65 per 1000 live births for boys) point out to the reality that women's second class status remains entrenched inspite of public policy measures and supportive legal frameworks.

Trafficking of women and children is another gross form of human rights violation. It deprives humans of their right to fundamental freedoms. Trafficked women and children, and those vulnerable to trafficking have a greater tendency to face risks, with fewer opportunities to avoid abusive situations, with marginalized access to justice and other resources for redress. Trafficking is often hidden in the veil of migration and so the abuse of rights is very difficult to identify in time.

Therefore, a significant number of trafficked women end up in prostitution and multiple conditions of exploitation as a consequence of coercion, deception and economic enslavement. Forced/bonded labour, domestic servitude, are among other major consequences of trafficking.

In the current scenario, the trend is that the age of trafficked victims is going down. A study by the Ministry of Women and Child Development commissioned in 2004 had estimated that around 2.8 million women are victims of trafficking, out of which 36% are children. Since trafficking is such a nuanced area of work the kinds of responses needed require multidisciplinary skills and multiple partnerships. Besides working on changing mind sets of communities and using communities aswatchdogs and agencies of change the work on trafficking requires that rescue and rehabilitation be done with a human rights approach through interventions that address the issues at all levels - at the level of the individual, at the level of the community, at the level of the state.

Various legal mechanisms have been put in place for addressing violence against women in the Indian Penal Code as well as through enactment of Special laws like the Protection of Women from Domestic Violence Act and Dowry Prohibition Act (DPA). However, enactment of law, though necessary as a first step to address violence against women and also to give them substantive equality, is not enough. It has to be effectively implemented and enforced. Further, victims and survivors of violence need services and support from the police, health, service providers, including legal aid and a sensitive judiciary.

❑❑❑

Chapter 5

Status and Contribution of Women in Agriculture and Allied Sectors

Multi-Dimensional Role of Women

i) **Agriculture** - sowing, transplanting, weeding, irrigation, fertilizer application, plant protection, harvesting, winnowing, storing, etc.

ii) **Domestic**- cooking, child rearing, water collection, fuel wood gathering, household maintenance, etc.

iii) **Allied sectors** - cattle management, fodder collection, milking, etc.

Rural women are mainly engaged in agricultural activities in three different ways depending on the socio-economic status of their family and regional factors. They work as (i) Paid Labourers (ii) Cultivator doing labour on their own land and (iii) Managers of certain aspects of agricultural production by way of labour supervision and participation in post harvest operations.

Social Dimensions of Women in India (2011 Census)

1.	Sex Ratio	940
2.	Sex Ratio in Rural Areas	947
3.	Literacy Rate in India	74%
4.	Females literacy rate	65.50%
	a) Rural female literacy rate	58.80%
	b) Urban female literacy rate	80%
6.	Female Life Expectancy (in the year 2006)	60 years
7.	Death Rates	
	Male	7.8
	Female	6.3

Economic Dimension of Women in India

Female work participation rates in 2009-10 in rural India have doubled when compared to those in urban India. In rural India, it is 26% and in urban India it is 14%.

Women Workers in Agriculture (Millions)

Agriculture	1981 Census	1991 Census	2011 Census
Cultivators	14.8	21.5	35.1
Agriculture Labourers	20.8	28.2	59.1

Distribution of Labour Force in India as per 2011 Census (Millions)

Item		2011
Total Population	Total Population	1210.6
	Male	623.1
	Female	587.4
Total Workers	Person	481.7
	Male	331.9
	Female	149.9
Total Marginal Workers	Person	119.3
	Male	58.7
	Female	60.6
Cultivators	Person	118.7
	Male	82.7
	Female	82.7
Agricultural Labourers	Person	144.3
	Male	82.7
	Female	61.6

(*Source* www.indiastat.com)

- Female population constitutes 49% of the total population and also 49% of the total rural population in India
- 31% of the total workers and 35% of the total rural workers in India are women
- But majority of the main workers are men with 75% share in total and 72% share in rural total main workers

- The percentage of females under the marginal work categories is more when compared to males
- Females constitute almost equal share in labour activities both at all India level and also in rural areas
- However, as cultivators, their role is far less when compared to male counterparts.

Wage Rates in Agriculture in India as per 2011 Census

Activity	Gender	Wage Rate (Rs)
Weeding	Male	98.96
	Female	81.1
Transplanting	Male	105.46
	Female	88.18
Harvesting	Male	106.87
	Female	88.69
Threshing	Male	108.3
	Female	86.93

- In all activities there is an average gender wage disparity, with women earning only 70 per cent of men's wage.
- Females are getting maximum wage rate for harvesting the agriculture produce.
- Wage rate of female varies from Rs 81 to Rs 89 and in case of male the wage rate varies from Rs 98 to Rs 108.

Percentage of Female in the Total Population and Workforce in India

Details	2001	2011
Population	48.3	48.5
Total Workers	31.6	31.1
Main Workers	23.2	24.6
Marginal Workers	60.3	50.8
Main and Marginal		
Cultivators	32.4	30.3
Agricultural Labour	46.6	42.7
House Hold Industry	49.3	46.7

Others	18.3	21.8
Agriculture	38.9	37.1
Non-agriculture	21.3	23.9
Main Workers		
Cultivators	24.4	23.8
Agricultural Labour	35.2	35.9
Household Industry	38.5	38.9
Others	15.2	18.3
Agriculture	28.5	29.5
Non-agriculture	17.2	19.7
Marginal Workers		
Cultivators	65.9	57.6
Agricultural Labour	63.0	52.7
Household Industry	80.7	62.8
Others	41.0	40.2
Agriculture	64.0	54.1
Non-agriculture	48.6	43.7

Source: **Census of India**

Education

- In 2012, the ratio of female to male primary school enrollment was 103%.
- In 2011, the ratio of female to male literacy was 91%.
- In 2012, the ratio of female to male secondary school enrollment was 95%.
- In 2011, 59% of women aged 15+ were literate compared to 79% of men.
- In 2010, women had on average 5 years of schooling compared to 8 years on average for men.

Health

- In 2013, there were 50,000 maternal deaths, representing 190 deaths for 100 000 births.
- In 2013, the adolescent fertility rate was 32 births per 1,000 adolescent girls (aged 15 to 19).

- There is no recent data available on the percentage of women with an unmet need for family planning.

Employment

- In 2012, 28% of the female working-age population was part of the labour force, while 82% of the male working-age population was.
- In 2012, women represented 24% of the total labour force.
- In 2010, 85% of women in the labour force were in vulnerable employment compared to 79% of men in the labour force.
- In 2012, 60% of women in the labour force were employed in the agricultural sector compared to 43% of men in the labour force.

Chapter 6

Gender Analysis

Gender analysis is the first and most critical step towards gender-responsive planning and programming. It involves the collection and analysis of sex disaggregated information. It examines the differences, commonalties and interactions between women and men. Gender analysis examines women and men specific activities, conditions, needs, access to and control over resources, access to development benefits and decision-making. It directs "WHO DOES WHAT" and provides us with precise information on gender perspective that is required for gender sensitization. The answers are based on facts regarding the patterns in the lives of women and men.

Gender Analysis: Is the methodology for collecting and processing information about gender.

The Main Goals of Gender Analysis

- To better understand our communities
- To promote gender equality through our work
- To expose the barriers of women's full participation and economic development.
- To help us find the best strategies and solutions and to address the different needs and dynamics of men and women living in a poverty.

When should we use Gender Analysis

- Evolution of the policy with principles
- Planning

- Programming
- Budgeting
- Implementation
- Monitoring and
- Evaluation

The Aim of using Gender Analysis in Agriculture is to:

- Improve overall project performance
- Overcome gender-based barriers
- Promote equal opportunities
- Increase both men's and women's participation.
- Ensure that new technologies will not have an adverse impact on women

To conduct a Gender Analysis, a core set of issues should be addressed. These are:

Women's and men's roles	Who does what, with what resources? Paying particular attention to variations within subgroups of women and men (e.g. Elderly women, adolescent girls, men from urban areas, etc). Typically, women perform three kinds of roles: - Productive roles (paid or not); - Reproductive roles (sustaining family living conditions and basic needs usually unpaid work) and - Community role.
Factors that shape gender roles and the gender division of work	Depending on the circumstances, traditions and institutions that shape gender roles represent constraints and / or opportunities for women and men. Understanding to what extent, and when, they are critical to designing the programs and projects suitableto the community
Access to and control overresources and	Not all men and women have the same access to and control over resources and

opportunities, and their systems of distribution	opportunities. Understanding the mechanisms and rules by which the resources and benefits are distributed is important to assess the situation of women vis-à-vis men (and vice versa) and determine the most effective entry points for action
Access to and participation in decision-making processes	Who decides? How are decisions taken concerning women's and men's lives and those of their families? Are women and men equally represented or given an opportunity to influence such processes?
Men's and women's practical and strategic needs and interests	Given their respective roles, who needs what for what purpose?

Gender Analysis of Activities

The format below can be used for analysis of roles played by men and women in different crop production activities, livestock rearing, household and off-farm production activities. Participation of both male and female members of the households should be ensured for analyzing the activities performed in each sector.

Crop Production

Crop: ....

Land: ....

Tasks	Males	No. of days	Females	No. of days
Task 1				
Task 2				
Task 3				

Livestock Production

Animal: ....

Tasks	Males	No. of days	Females	No. of days
Task 1				
Task 2				
Task 3				
Task 4				

Access & Control Profile

The following table can be used in a focused group discussion for understanding the access and control profile of men and women over different resources, in a household and community.

Resources	Access		Control		Comment
Land	Women	Men	Women	Men	
Equipment					
Labour					
Cash, Gold / Jewellery					
Employment Opportunities					
Education/Trainings					
Others					

The information generated using the above tables will help the extension functionaries and the community to understand the existing scenario of gender division of work/roles, work load and the time spent on different activities. Further, it will be helpful to plan the extension activities as per the needs of the farming community.

In socio-cultural context especially in India, family, community, and the existence of different forms of institutions like caste or cultural community, are the institutions defining the social status and gender definitions for both women and men. When we try to make out a gender analysis in this context, it's more identifying how the role of men and women is constructed in terms of childhood nurturing, marriage, ceremonies, family, religious rituals, etc.

Similarly, the gender roles may vary from caste to caste, family to family like nuclear family vis- a- vis joint family. Hence, here the gender analysis is more of subjective in nature.

Therefore, the method of Gender Analysis has to consider that:

- All knowledge and scientific data is available within the target group of people.
- Participants play equal role with the analyst in terms of defining terms, relationships and situations.
- The process of analysis empowers the people involved through real life stories and leads to transformation.

Gender analysis is a way of seeing/ analyzing problems, situations and solutions with awareness of gender relations and in order to identify gender issues. The key elements of a gender analysis should include identification of similarities and differences between men and women and amongst women. These relate to work, resources, responsibilities and powers; assessment of how gender relations have an impact on opportunities, needs, incentives and rewards; assessment of capacity of intervening institutions to promote gender equality goals; estimation of the potential obstacles and resistances to initiatives to promote gender equality and development of strategies to counter these resistances. It is important to understand practical needs and strategic needs of women.

Practical needs can be defined as women's needs that do not question her subordination or traditional roles. These are often to do with the practical needs of day to day living such as housing, water; healthcare and employment.

In contrast **Strategic Gender** interests are the needs of women identified on the basis of their subordinate position in society. They may vary from context to context and may include issues, such as legal rights to freedom from domestic violence, rights to land and property, sexual rights, unequal divisions of labour in the family, etc. These are often less obvious and have long term implications.

The United Nations Development Programme synthesizes the steps of gender analysis in this way:

(1) Sex-Disaggregated Data + (2) Analysis + (3) A Gender Perspective = Gender Analysis

In general, all gender analysis approaches should examine representation of men and women in a particular sector (e.g., education, health, or economic growth), how resources are distributed, and why these differences exist. It may also be useful to examine the historical and contemporary social context relevant to the specific sector to understand gender differences.

Analysis of Sex-Disaggregated Data and Information

Review any sex-disaggregated data relevant to the specific technical sector's proposed project or activity. This data can be found in published sources from the World Bank and other international agencies, from gender experts, stakeholder consultations, or collected through other vehicles, such as surveys or focus groups.

Assessment of Roles and Responsibilities/Division of Labour

- Consider men's and women's roles and responsibilities in relation to formal and informal employment, community participation, local/community politics, family and household roles, etc.
- Where is there greater participation of women or men? What are the reasons for this?
- What are the barriers to an equitable division of labour and equal opportunity?
- Where do opportunities or entry points exist to encourage an equitable division of labour and equal opportunity?
- What different roles do boys and girls play within the household and community? How might this affect their access to education, health care, etc.?

Consideration of Access to and Control Over Resources

Who has access to and control of resources and services?

It is important to distinguish between the concepts of access and control when examining how resources are allocated between women and men. Access means a person or group can use a resource (e.g., land to grow crops), while control means that a person or group can make decisions about who uses or disposes of a resource (e.g., selling land and gaining the profits).

Examination of Decision-Making Patterns

In what decision-making do men and women participate? Analysis can include the national, community, and household level.

There may be formal constraints on women to participate in decision-making, while in others, the impediments may be de facto. Analyze and discuss how women and men are actually participating in decision-making. Such an analysis should include not only who is present when decisions are made but also who has actual power over the outcomes of the decision-making process.

Gender Analysis Frameworks

Gender analysis being a type of socio-economic analysis, it is intended as a tool to illuminate the links between the existing gender relations in a particular society and development problems it needs addressed. Generally, Gender Analysis is made to answer the following two questions:

1. How will gender relations affect the achievement of sustainable results? and
2. How will proposed results affect the relative status of men and women?

There are several frameworks and methodologies to conduct a Gender analysis. Each model has its strengths and weaknesses. Some are useful for micro-planning and give greater importance to gender roles (Harvard Framework), while others emphasize the enquiry into social relations. Some have been designed to exclusively look at women's empowerment (Longwe Framework).

Commonly Referenced Gender Analysis Frameworks

This section introduces five of the commonly used gender analysis frameworks:

- The Harvard Analytical Framework, also known as the Gender Roles Framework
- The Moser Gender Planning Framework
- The Gender Analysis Matrix (GAM)
- The Women's Empowerment Framework (WEP), and
- The Social Relations Approach

1. Harvard Analytical Framework

The Harvard Analytical Framework (sometimes referred to as the "**Gender Roles Framework**" or the "**Gender Analysis Framework**") was developed by researchers at the Harvard Institute of International Development (HIID) in collaboration with USAIDs Office of Women in Development. It represents one of the earliest efforts to systematize attention to both women and men and their different positions in society. It is based upon the position that allocating resources to women as well as men in development efforts makes economic sense and will make development itself more efficient - a position labeled as the "efficiency approach."

Key to the Harvard Analytical Framework is adequate data collection at the individual and household level, and adapts well to agricultural and other rural production systems. Data is collected on men's and women's activities which are identified as either "reproductive" or "productive" types, and is then considered according to how those activities reflect access to and control over income and resources, thereby "highlighting the incentives and constraints under which men and women work in order to anticipate how projects will impact their productive and reproductive activities as well as the responsibilities of other household members."

The Objective of this framework is:

- To demonstrate that there is an economic rationale for investing in women as well as men.
- To assist planners design more efficient projects.
- To emphasise importance of good information as basis for efficient/effective projects.
- To map the work of women and men in the community and highlight differences.

Features

The framework consists of a matrix for collecting data at the micro (community and household) level. It has four inter-related components:

- The activity profile, which answers the question, "who does what?", including gender, age, time spent and location of the activity

- The access and control profile, which identifies the resources used to carry out the work identified in the activity profile, and access to and control over their use, by gender
- The analysis of influencing factors, which charts factors that influence gender differences in the above two profiles
- The project cycle analysis, which examines a project or intervention in light of gender-disaggregated information

The framework also contains a series of checklists consisting of key questions to ask at each stage of the project cycle: identification, design, implementation, and evaluation.

Benefits of the framework

- Best suited for project planning, rather than programme or policy planning
- As a gender-neutral entry point when raising gender issues with constituents resistant to considering gender relations and power dynamics
- For baseline data collection
- In conjunction with Moser's framework, to draw in the idea of strategic gender needs.

Strengths and Weakness of Harvard framework

Strengths	Limitations
• Practical and easy to apply • Helps collect and organize information about gendered division of work • Helps make women's work visible • Distinguishes between resource access and control • Adaptable to a variety of settings and situations • Non-threatening inquiry about women or men's activities because it relies on "facts"	• Focus on material resources rather than strategic gender needs (e.g., increasing women's access to decision-making) • Tends to oversimplify and homogenize inequalities (e.g., differences between mothers-in-law and daughters-in-law) • Ignores underlying inequalities, such as class, race, or ethnicity (e.g., poor women and wealthy women will not necessarily perform the same kind of agricultural or household work)

Harvard Analytical Framework is a simple and practical toolset to identify the type and amount of work men and women do in a household, farm, or community. It also helps to document the differences in the gendered access and control of resources, such as land, water, seeds or extension information. The information gathered from the toolset can be used by extension agents to clarify what men and women do, and then tailor programs to meet the specific needs and interests of farmers or community members, especially women. This toolset can be easily adapted for use in either agricultural or livestock (or combined) production systems.

The value of the toolset depends on the collection of sufficiently detailed data at the individual, household or community level. One tip for data collection is to hold participatory meetings with farmers or community members. These meetings allow everybody to contribute to the discussion and help men and women map out "facts" and understand who is doing what, when and how. It may be important to collect this information in separate groups of men and women if the women do not feel comfortable speaking in mixed groups, and then have male and female groups come together to share findings. The data is gathered through three inter-related tools: an activity profile, an access and control profile, and a list of influencing factors. Each tool contains a series of checklists of key questions every extension agent can use.

The Activity Profile: Inquiries about who does what for all relevant farm, home, and community tasks:

- ❒ Who does what?
- ❒ When?
- ❒ Where?

Activity Profile Tool (example)

	Women	Men	Girls	Boys
Productive Activities				
Agriculture				
• activity 1				
• activity 2				
Income generating:				
• activity 1				
• activity 2				
Employment				
Reproductive Activities				
Water				
Fuel				
Cooking				
Childcare				
Health				
Cleaning				
Repair				
Community Involvement				
Meetings				
Wedding				
Helping others in time of need				
Labour on communal projects				

The Access and Control Profile: Clarifies and documents who has access to resources and who controls their use in relation to the tasks identified in the Activity Profile

- ❐ Who has access to and control over resources (for example, land, labor, extension services)?
- ❐ Who has access to and control over benefits (education or health services)?

Access & Control Profile Tool (example)

	Access		Control	
	Women	**Men**	**Girls**	**Boys**
Resources				
Land				
Water				
Seed				
Labour				
Extension				
Etc.				
Benefits				
Assets				
Income				
Education				
Political power/ prestige etc.				

Influencing Factors: Identifies factors that shape differences between men and women:

- What are political, economic, or cultural factors affecting the gender differences identified in the above two profiles?
- What are the past and present influences?
- What are the opportunities and constraints?

Influencing Factors Tool (example)

	Women	Men	Girls	Boys
Factors				
- Community norms				
- Social hierarchy				
- Institutional				
- Economic				
- Political				

Once the data have been analyzed extension agents can use findings from the tool to plan their extension programs, including:

- Tailoring training sessions to women needs (e.g., hold meetings at time or locations convenient to women)
- Adjusting extension program agenda (e.g., include crops grown by women for household consumption rather than markets)
- Planning new extension activities (e.g., increase demonstration plots on women's fields)

2. Gender Planning Framework (Caroline Moser)

This framework, developed by ***Caroline Moser,*** links the examination of women's roles to the larger development planning process. The framework introduces the idea of women's "three roles" in production, reproduction, and community management, and the implication that these roles have for women's participation in the development process. In making these links, both between women and the community, and between gender planning and development planning more broadly, Moser's framework encompasses both the technical and political aspects of gender integration into development.

The Moser framework follows the Gender and Development approach in emphasizing the importance of gender relations. As with the WID-based Harvard Analytical Framework, it includes collection of quantitative empirical facts. Going further, it investigates the reasons and processes that lead to conventions of access and control. The Moser Framework includes gender roles identification, gender needs assessment, disaggregating control of resources and decision making within the household, planning for balancing the triple role, distinguishing between different aims in interventions and involving women and gender-aware organizations in planning. The framework acknowledges a political element to gender planning, and assumes that the process will have to deal with conflicts.

The objective of this framework is:

- Focus on strategic gender needs and concentrates on gender inequalities and how to address these at programme and policy level.

Components of the model

The framework is composed of several components (or tools). In the first, the triple roles of women are identified by mapping the activities of household members (including children) over the course of twenty-four hours.

1. Gender roles identification - focus on triple roles of women (productive, reproductive and community)
   - **Reproductive Roles**: Childbearing and rearing, domestic tasks that guarantee the maintenance and reproduction of the current and future work force (e.g., cooking, cleaning, etc.)
   - **Productive Roles**: Work done for remuneration, in cash or kind. (e.g., wage labor, farming, crafts, etc.)
   - **Community Management Roles**: Work that supports collective consumption and maintenance of community resources (e.g., local government, irrigation systems management, education, etc.)
2. The second component identifies and assesses gender needs, distinguishing between practical needs (to address inadequate living conditions) and strategic needs (for power and control to achieve gender equality).
3. The third component, or tool, disaggregates information about access to and control over resources within the household by sex: who makes decisions about the use of different assets.
4. The fourth component identifies how women manage their various roles, and seeks to clarify how planned interventions will affect each one.

5. Finally, the WID/GAD policy matrix evaluates how different planning approaches (welfare, equity, anti-poverty, efficiency, and empowerment) have addressed the triple roles and women's practical and strategic needs.

Best Suited

- For planning at all levels from policies to projects.
- In conjunction with the Harvard Framework

Strengths and Limitations

Strengths	Limitations
• Assumes planning exists to challenge unequal gender relations and support women's empowerment.	• Framework does not mention other inequalities like class, race and ethnicity.
• Makes ALL work visible through concept of triple roles.	• Framework is static and does not examine change over time.
• Alerts planners to interrelationships of triple roles.	• Looks at separate, rather than inter-related activities of women and men.
• Recognises institutional and political resistance to transforming gender relations.	• Strict division of practical and strategic needs not always helpful in practice.
• Distinguishes between practical gender needs (those that relate to women's daily life) and strategic gender needs (those that potentially transform the current situation)	• Strategic needs of men not addressed.

Using this Framework in Agricultural Extension and Advisory Services

1. With the participation of men and women together or in separate groups, identify and map out all the activities of men and women in the household over a 24-hours period using the "triple roles" framework. Include girls and boys in the mapping exercise to obtain a fuller picture.

2. Clarify the needs of men and women by distinguishing between their practical needs (e.g., is there food to cook for dinner?) and

strategic needs (e.g., access of women to decision-making or women's control of their time).

3. Discuss results of 1 and 2 with the group.
4. Use this information to assess how men and women manage their various roles and deciding on the timing of extension programs that are suitable to women's available time and does not create additional burden on women's responsibilities in the household.
5. Identify the type of extension interventions that you can tailor to men and women's roles and their needs.
6. Involve men and women during the planning process to design and offer gender-effective extension programs.

3. Gender Analysis Matrix (GAM)

The Gender Analysis Matrix (GAM), developed by ***Rani Parker,*** is a tool that uses participatory approaches to identify how a particular agricultural production, processing or marketing practice impacts men and women differently in the community. Its community-focused approach enables participants to analyze differences between men and women's culturally-defined practices in the community, and challenge their assumptions about gender roles.

For example, a community could evaluate the labour practices of men and women, and assess how they impact men and women's wellbeing, time use or earning potential. If they identify a particular activity that discriminates against women, the community may decide to stop such a practice and create a more equitable environment.

The benefits of using GAM are based on its community-based techniques to collect and analyze information on the spot. The targeted community self-identifies problems and finds equitable solutions prompting transformative change. The use of participatory approach, where a group of men or women, or mixed group of men and women identify and propose culturally-validated changes at the community level.

Basics of GAM

To ensure effective gender analysis process with the use of GAM, it is important to understand that:

- Participating men and women possess all necessary information about themselves and their community.
- Application of GAM doesn't require the community to have gender expertise, but extension agents facilitating the process should possess an understanding of gender analysis.
- Application of GAM facilitates transformative change in the community as the community self-identifies problems and consensually validates its own solutions.

Parker (1993) describes GAM using What, Why, Who, When and How

- **WHAT**: A tool for gender analysis of development (program/ project) at community level.
- **WHY**: To determine differing impacts of interventions on women & men.
- **WHO**: Done by a group within the community; includes women & men in equal number
- **WHEN**:
  - At the ***planning*** stage →determine potential gender effects
  - At the ***design*** stage → gender considerations may change the design
  - During ***monitoring*** stage →address broader program impacts

The framework

The GAM is based on the following principles:

- All requisite knowledge for gender analysis exists among the people whose lives are the subject of the analysis.
- Gender analysis does not require the technical expertise of those outside the community, except as facilitators.
- Gender analysis cannot promote transformation unless it is carried out by the people being analysed.

GAM Matrix

Intervention	Labour	Time	Resources	Culture
Men				
Women				
Household				
Community				

GAM Tool 1: Level of Analysis

GAM allows analysis of an intervention at four levels: men, women, households, and community. The levels of analysis appear vertically on the matrix:

- **Men:** Represent men of all ages who are in the target group or all men in the community.
- **Women:** Represent women of all ages who are in the target group or all women in the community.
- **Household:** Represents all women, men, and children living under one roof.
- **Community:** Represents everyone in the community.

GAM Tool 2: Impact Analysis

GAM examines impact on four areas, which appear horizontally on the matrix:

- **Labour:** Captures changes in tasks (do women take over men's tasks in the field), the level of skill (formal education, training) required, the number of people involved in this activity and the demand for additional labour.
- **Time:** Captures changes in time requirements to complete specific tasks.
- **Resources:** Captures changes in access to resources (income, land, extension information) and the extent of control over resources (increase or decrease) as a result of an intervention.
- **Socio-cultural factors:** Captures changes in gender roles or status as a result of an intervention.

Strengths and Limitations

Strengths	Limitations
• Designed specifically for community-based development workers • Simple and systematic uses familiar categories and concepts • Transformatory as well as technical • Fosters 'bottom-up analysis through community participation • Considers gender relations between women and men, as well as examining what each category experiences separately • Levels of analysis can be added to in order to suit particularly interventions • Includes intangible resources • Can be used to capture changes over time • Helps anticipate resistance, and encourages consideration of what support should be offered for those at risk • Includes men as gendered beings, so can be used in interventions that target men • Can be used for participatory impact assessment • Quick data gathering	• Needs a good facilitator • Some factors can get lost because categories have many aspects • Requires careful repetition in order to consider change over time • Does not seek out the most vulnerable community members • Excludes macro- and institutional analysis • Difficulties defining a community • Subordination is often no explicit • Risk of misleading outcomes due to power relations between funders and community members

4. Social Relations Framework: Naila Kabeer

The social relations framework originated with academics led by Naila Kabeer at the Institute of Development Studies at Sussex, UK. It is outlined in her book, *Reversed Realities: Gender Hierarchies in Development*, Verso, 1994.

AIMS of the Framework

- To analyse existing gender inequalities in the distribution of resources, responsibilities, and power
- To analyse relationships between people, their relationship to resources and activities, and how they are reworked through institutions
- To emphasise human well-being as the final goal of development

Features

Five essential concepts

1. Development is increasing human well being (survival, security autonomy)
2. Social relationship analysis. The way people are positioned in relation to tangible and intangible resources.
3. Institutional analysis Key institutions; state, market, legal, family/kinship. Aspects of institutions; rules, activities, resources, people, power.
4. Institutional gender policy analysis
5. Analysis of underlying and structural causes and the effects of these.

Gender analysis, therefore, entails looking at how institutions create and reproduce inequalities. There are four key institutional sites: the state, the market, the community and family/kinship.

Institutional location	Organisational/structural form
State	legal, military, administrative organisations
Market	firms, financial corporations, farming enterprises, multinationals
Community	village tribunals, voluntary associations, informal networks, patron-client relationships, NGOs
Family/kinship	household, extended families, lineage groupings

Five dimensions of institutional social relationships are especially relevant for gender analysis

- *Rules*, or how things get done; do they enable or constrain? Rules may be written or unwritten, formal or informal
- *Activities*, or who does what, who gets what, and who can claim what. Activities may be productive, regulative, or distributive
- *Resources*, or what is used and what is produced, including human (labour, education), material (food, assets, capital), or intangible resources (goodwill, information, networks)
- *People*, or who is in, who is out and who does what. Institutions are selective in the way they include or exclude people, assign them resources and responsibilities, and position them in the hierarchy
- *Power*, or who decides, and whose interests are served.

Naila Kabeer classifies development policies as follows:

1. **Gender-blind**
    - Do not distinguish between men and women
    - Incorporate existing biases
    - Tend to exclude women
2. **Gender-aware**
    - Recognise differences among men and women's needs and priorities

Gender aware policies may be of three types

Gender-Neutral	• In light of gender differences, target delivery to men and women's practical gender needs • Work within existing gender division of resources and responsibilities
Gender-Specific	• In light of gender differences, respond to the practical needs of men or women specifically • Work within existing gender division of resources and responsibilities

Gender Redistributive	• Intend to transform existing gender relations to create a more balanced relationship • May target both men and women, or one specifically • Work on practical gender needs in a transformatory way • Work on strategic gender needs

Finally, the social relations framework analyse immediate, *underlying* and *structural* causes of specific gender issues and their effects, as shown in the table below:

Analysis of causes and effects

Long-term effects

Intermediate effects

Immediate effects

The Core Problem

Immediate causes at level of

- household
- community
- market
- state

Intermediate causes at level of

- household
- community
- market
- state

Structural causes at level of

- household
- community
- market
- state

Uses of the framework

- ❐ Can be used from project to policy level planning, even on an international basis

Strengths and Limitations

Strengths	Limitations
• Presents a broader picture of poverty. • Conceptualises gender as central to development thinking not an add-on. • Used at different levels for planning and policy development. • Links micro and macro analysis. • Centres analysis on institutions and highlights their political aspects. • Highlights interactions between inequalities - race, class, and ethnicity. • Dynamic - works to uncover processes of impoverishment and empowerment	• Can appear to be complicated. • Since it looks at all inequalities - it can subsume gender into other analytical categories. • Can overlook the potential for people to effect change. • May give an overwhelming impression of large institutions

Womens Empowerment Framework

The Women's Empowerment Framework was developed by ***Sara Hlupekile Longwe,*** a gender expert from Lusaka, Zambia. Her model is explicitly political, arguing that women's poverty is the consequence of oppression and exploitation (rather than lack of productivity), and that to reduce poverty women must be empowered.

Objective of the Women's Empowerment Framework is

- ❐ To achieve women's empowerment by enabling women to achieve equal control over factors of production and participate equally in the development process.

Features

Framework introduces five hierarchical levels of equality (the higher you go the more empowered you are):

1. **Control** - equal control over in decision-making over factors of production.
2. **Participation** - equal participation in decision-making processes related to policy-making, planning and administration.
3. **Conscientisation** - attaining equal understanding of gender roles and a gender division of labour that is fair and agreeable.
4. **Access** - equal access to the factors of production by removing discriminatory provisions in the laws.
5. **Welfare** - having equal access to material welfare (food, income, medical care).

Framework distinguishes between women's issues and concerns as well as identifying three levels of recognition of women's issues in project design.

Longwe also distinguishes between

Women's issues	Which pertain to equality with men in any social or economic role and involving any of the levels of equality
Women's concerns	Which pertain to women's traditional and subordinate, sex-stereotyped gender roles

The women's empowerment framework identifies three levels of recognition of women's issues in project design:

Negative level	Where project objectives are silent about women's issues. Experience suggests that women are likely to be left worse off by such a project
Neutral level	Where the project objectives recognise women's issues but concern remains neutral or conservative, merely ensuring that women are not left worse off than before
Positive level	Where project objectives are positively concerned with women's issues and with improving the position of women relative to men

The framework can be used to produce profiles as below:

Levels of Recognition	Levels of Equality		
	Negative	Neutral	Positive
Control			
Participation			
Conscientisation			
Access			
Welfare			

Women's Empowerment Framework for a Country Programme

		Level of concern with women's development				
Sector	Project	Welfare	Access	Conscien-tisation	Partici-pation	Control
Agriculture						
Education						
Industry						
Women's projects						

Best suited in

- Useful across micro (project) and macro (country strategy) levels of analysis.
- Useful where focus is specifically on empowerment of women.

Strengths and Limitations

Strengths	Limitations
• Framework can be used to prepare profiles of levels of recognition as well as profiles of analysis of levels of equality across sectors. • Develops notion of practical and strategic gender needs into progressive hierarchy. • Articulates empowerment as essential element of development. • Enables assessment of interventions based on grounds of empowerment. • Has a strong political perspective-aims to change attitudes.	• Assumption that a level of equality is strictly hierarchical is questionable. • Framework profiles are static and do not take account of changes over time. • Focus on gender equality only takes no account of inter-relationships between rights and responsibilities. • Ignores other forms of inequality.

❑❑❑

Chapter 7

Gender Impact Assessment

Gender impact assessment has been defined as an ex ante evaluation, analysis or assessment of a law, policy or programme that makes it possible to identify, in a preventative way, the likelihood of a given decision having negative consequences for the state of equality between women and men.

The central question of the gender impact assessment is: Does a law, policy or programme reduce, maintain or increase the gender inequalities between women and men?

The European Commission defines gender impact assessment as follows:

"Gender impact assessment is the process of comparing and assessing, according to gender relevant criteria, the current situation and trend with the expected development resulting from the introduction of the proposed policy."

"Gender impact assessment is the estimation of the different effects (positive, negative or neutral) of any policy or activity implemented to specific items in terms of gender equality."

The assessment involves a dual-pronged approach: the current gender-related position in relation to the policy under consideration, and the projected impacts on women and men once the policy has been implemented. It is important that the assessment is structured, i.e. systematic, analytical and documented.

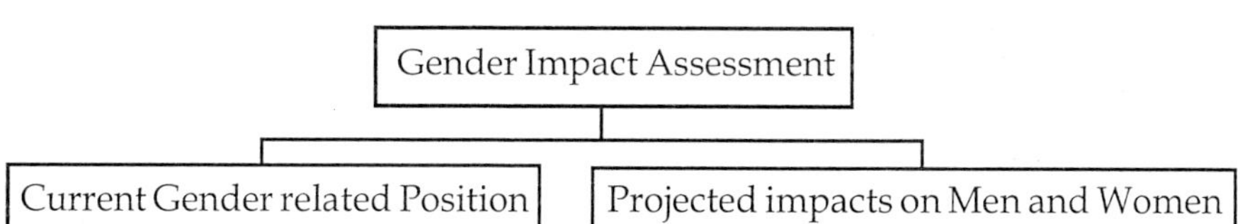

The final aim of the gender impact assessment is to improve the design and the planning of the policy under consideration, in order to prevent a negative impact on gender equality and to strengthen gender equality through better designed, transformative legislation and policies. A primary objective is to adapt the policy to make sure that any discriminatory effects are either removed or mitigated. Beyond avoiding negative effects, a GIA can also be used in a more transformative way as a tool for defining gender equality objectives and formulating the policy so as to proactively promote gender equality.

When is gender impact assessment needed?

The Council of the European Union, in its conclusions from 2006, noted that despite some progress towards gender mainstreaming in Member States, gender impact assessment still needs to either be put in place or reinforced. The Council urged in particular to improve and strengthen the development and regular use of gender impact assessment when drafting:

- Legislation
- Policies
- Programmes
- Projects

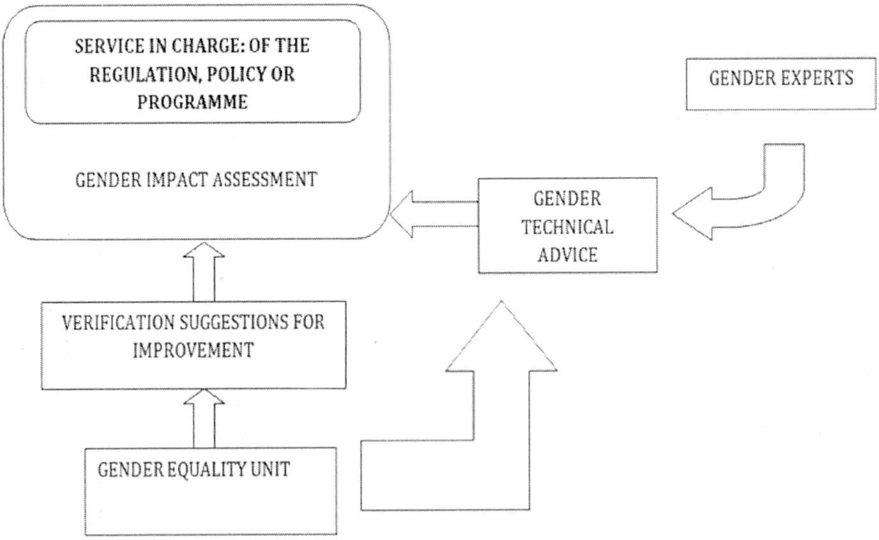

Steps to carry out gender impact assessment

Some defined steps are needed to carry out a gender impact assessment, even though the number of steps defined may vary from one context to another, depending on the approach taken. However, the stages of the GIA process are always the same.

- Step 1 : Definition of Policy Purposes
- Step 2 : Checking Gender Relevance
- Step 3 : Gender Sensitive Analysis
- Step 4 : Weighing the Gender Impact
- Step 5 : Findings and Proposals

Step 1: Definition of the policy purpose

The first step is to define the purpose of the planned policy, law or programme and show how it connects with gender equality. The following questions could be asked:

- What social issue is being addressed by this policy/ legislative intervention?
- Why is this intervention being considered for this particular situation?
- Is the intervention intended to contribute to gender equality?
- How is the intervention intended to contribute to gender equality?
- What are the existing gender equality objectives in this field?

Step 2: Checking gender relevance

Once the law, policy or programme has been contextualised, it is necessary to determine its gender relevance. This involves analysing whether or not it is likely to impact on gender equality. To do this, the following elements must be taken into account:

The proposed initiative will be applicable and therefore, have to assess the gender impact whenever these two conditions are met:

1. **Target group:** The ultimate target group is people, considering both individuals and legal entities.
2. **Impact on the target group:** The proposed initiative affects women and men regarding their access to and/or control of resources. Consequently, it has an impact on the social situation or position of women and men, by either improving or harming it.

The impact of any law, policy or programme on the target group may be direct or indirect:

1. **Direct impact:** When regulating or affecting people's access to resources (grants, jobs, composition of committees, etc.). As a result, it has a direct and immediate effect on the status and position of women and men. Example: The launch of an initiative that aims to promote the employment of certain groups of people (the young, people over a certain age, etc.)
2. **Indirect impact:** When regulating or planning measures affect the means of provision of certain resources or services (procedure to qualify companies, regulation of environmental quality management of certain activities and facilities, incentives for certain projects, etc.), behind which there are people (managers, workers, users, etc.) as ultimate beneficiaries. Even though the policy is not directly targeted at them, they can be affected by it. Example: A rule, policy or programme that seeks technological development within companies. Although the target groups are enterprises, these companies are formed by women and men, both as owners and as workers, with different positions and usually in unequal situations, due to horizontal and vertical gender segregation. Therefore, the initiative concerned can help to strengthen, maintain or reduce these inequalities.

Step 3: Gender-sensitive analysis

All regulations, policies and programmes relevant to gender affect the living conditions and access to resources for women and men.

This effect is known as 'gender impact'. An assessment of the gender impact of any initiative involves a process of analysis to determine the impact (tangible results) that the intervention could have on the effective equality of women and men.

On one hand, such analysis should try to comprehend the present situation for the groups concerned and understand how this situation could evolve without public intervention. On the other hand, the analysis should measure, as much as possible, how the planned intervention is expected to change the existing situation. In order to guide the analysis, the following recommendations can be taken into account:

1. **Analyse the situation of women and men in the field**

- Collect information and data about the existing situation for women and men in the areas concerned by the planned policy or legislative measure, looking for sex-disaggregated statistics.
- Complement quantitative information with qualitative insights, for example from studies or consultations, and combine various sources (statistical offices, academic works, policy reports) to gain a deeper understanding.

2. **Identify existing gender inequalities**

Take into consideration the views of those concerned (primarily the groups that are directly targeted). The following key questions could be used:

- What are their expectations and needs?
- Are these different for women and men?
- Is the planned intervention addressing the needs of both women and men, taking into account their different interests, roles and positions?
- How can the contribution to the needs of women and men be strengthened?

Step 4: Weighing the gender impact

In this phase it is important to establish how the policy or legislative measure will contribute to gender equality, as well as to assess the foreseen impact on gender relations. Once the effects of the proposed policy have been identified, they should be 'measured', taking into consideration the expected trends as previously identified.

To do this, it may be useful to take into account the following criteria so as to weigh the positive, neutral or negative gender impact of any initiative:

1. **Participation of women and men** - The impact will be considered positive when a significant increase in the representation of the under-represented sex in the area is envisaged. In this sense, the expected result should be a balanced representation of women and men (no less than 40 % for each sex) or, temporarily, at least in proportion to their overall presence in the area. Fostering gender equality implies promoting the participation of women in the public sphere (politics, economy, employment, culture, etc.), and increasing men's participation in domestic duties and care work. Finally, it is necessary to highlight one exception to these general rules. When talking about women's specific organisations, spaces or facilities to foster their empowerment and social involvement, it is paramount to understand that participation only by women is a relevant and accepted strategy to promote gender equality.
2. **Access to and control of resources** - The aim is to analyse the access of women and men to essential resources, such as education, employment, careers, health, time, money, power, information, new technologies, etc. The impact will be considered positive when the elimination of existing gender gaps, or at least a significant reduction of them, is foreseen. This reduction should always be aimed at increasing the physical, emotional and economic empowerment of women. In the case of men, the focus should be to increase their involvement in housework and care work.
3. **Gender-based social norms and values** - The aim is to evaluate how the project will impact on gender-based social norms and gender roles and how it will contribute to promoting the equal social value of women and men, femininity and masculinity. The impact will be considered positive when:
   - Mechanisms or structures that help to reproduce gender inequalities are modified. This means acting on the sexual division of labour, on the organisation of private life, or on the organisation of citizenship.
   - In the first instance, progress in eliminating gender stereotypes is made.

Step 5: Findings and proposals for improvement

Conclusions must be formulated in terms of the impacts on women and men within the target group(s). Furthermore, proposals should be put forward that will promote gender equality in response to the existing situation. To do this, one should identify how gender equality could be strengthened in the different parts of the draft initiative, taking into account the conclusions reached:

1. Gender equality should be mentioned in the preamble to the law or in the definition of the context of the policy or programme as well as in the legal framework referring to it.
2. Gender equality should also be mentioned in the objective so as to set a starting point and encourage the commitment of all stakeholders involved in the development of the initiative.
3. Actions to reduce imbalances and inequalities and measures to promote gender equality should be included among the different aspects of the law, policy or programme. These measures could be related to:
   - Promoting the access of women in sectors where they are under-represented
   - Promoting the access of women to decision making
   - Preventing gender-based violence
   - Promoting proactive actions to eradicate gender based violence
   - Eliminating gender stereotypes and roles

Another important component in this last step is the identification of indicators to allow for monitoring and evaluation of the actual outcomes. To do this, the following aspects should be taken into account:

- It is important to assess what indicators are currently available and how useful they are in tracking progress on gender equality. Alternatively, it may be necessary to set new indicators based on pertinent sex-disaggregated data.
- Where lack of information curtails a proper analysis and thus produces inconclusive findings, it is important that this is explicitly mentioned in the GIA report. Making data and information gaps known allows action to be taken by those responsible to remedy this problem.

Impact of the gender impact assessment

The implementation of an effective GIA process will result in an honest appraisal of the inequalities in the lives that women and men lead, a recognition of the consequences of those inequalities and, subsequently, a tailor-made response in policies and practices. Ultimately, the process will be judged on its outcomes. The opportunities presented must be accompanied by the resources and political will to make them materialise.

❑❑❑

Chapter 8

Gender Mainstreaming

Gender mainstreaming first came up as a concept in 1985 at the United Nations Third World Conference on Women in Nairobi. A decade later, in 1995, the Platform for Action adopted at the Fourth World Conference on Women in Beijing explicitly called upon governments and other actors to promote gender mainstreaming.

Gender mainstreaming is the process of assessing the implications for women and men of any planned action, including legislation, policies or programmes, in all areas and at all levels.

It is a strategy for making women as well as men concerns and experiences an integral dimension of the design, implementation, monitoring and evaluation of policies and programmes in all political, economic and societal spheres so that women and men benefit equally and inequality is not perpetuated. The ultimate goal is to achieve gender equality.

The concept of bringing gender issues into the mainstream of society was clearly established as a global strategy for promoting gender equality in the Platform for Action adopted at the United Nations Fourth World Conference on Women, held in Beijing (China) in 1995. It highlighted the necessity to ensure that gender equality is a primary goal in all area(s) of social and economic development.

In July 1997, the United Nations Economic and Social Council (ECOSOC) defined the concept of gender mainstreaming as follows:

"Mainstreaming a gender perspective is the process of assessing the implications for women and men of any planned action, including legislation, policies or programmes, in any area and at all levels. It is a strategy for making the concerns and experiences of women as well as of men an integral

part of the design, implementation, monitoring and evaluation of policies and programmes in all political, economic and societal spheres, so that women and men benefit equally, and inequality is not perpetuated. The ultimate goal of mainstreaming is to achieve gender equality."

Mainstreaming includes gender-specific activities and affirmative action, whenever women or men are in a particularly disadvantageous position. Gender-specific interventions can target women exclusively, men and women together, or only men, to enable them to participate in and benefit equally from development efforts. These are necessary temporary measures designed to combat the direct and indirect consequences of past discrimination.

Transformation by Mainstreaming

Mainstreaming is not about adding a "woman's component" or even a "gender equality component" into an existing activity. It goes beyond increasing women's participation; it means bringing the experience, knowledge, and interests of women and men to bear on the development agenda. It may entail identifying the need for changes in that agenda. It may require changes in goals, strategies, and actions so that both women and men can influence, participate in, and benefit from development processes. The goal of mainstreaming gender equality is thus the transformation of unequal social and institutional structures into equal and just structures for both men and women.

How does Gender Mainstreaming work?

A political commitment for gender equality and a compatible legal framework are the basic conditions for the development of a successful gender mainstreaming strategy. In addition to concrete objectives and targets in the strategy, gender mainstreaming requires a clear action plan. Such plan should take into account the context, satisfy the necessary conditions, cover all the relevant dimensions, foresee the use of concrete methods and tools, set out the responsibilities and make sure that the necessary competences exist to achieve the anticipated results within a planned timeframe.

Dimensions of gender mainstreaming

Gender mainstreaming requires both integrating a gender perspective to the content of the different policies, and addressing the issue of representation of women and men in the given policy area.

Both dimensions - *gender representation* and *gender responsive* content - need to be taken into consideration in all phases of the policy-making process.

Enabling conditions for gender mainstreaming

An effective implementation of gender mainstreaming requires preparation and organisation. People in decision-making positions can make a particular difference here, as they have more power to introduce changes.

Key elements to consider are:

- **Preparation**: set up a plan for the implementation of gender mainstreaming, define steps and milestones, assign tasks and responsibilities, formalise and communicate the plan.
- **Resources**: sufficient resources need to be made available; effective gender mainstreaming requires budget and time. Think about resources for awareness-raising and capacity-building initiatives. The use of special (external) expertise might also be considered.
- **Stakeholder involvement**: close liaison with all policy stakeholders is essential throughout the policy cycle to take on board the concerns, expectations, and views of the target groups. It is recommended to cement opportunities and structures for stakeholder involvement and consultations into the policy process.
- **Monitoring and evaluation**: set in place accountability mechanisms to ensure an adequate follow-up of implementation and progress. Foresee regular reporting and share results.
- **Knowledge generation**: building up knowledge on gender equality and good practices in gender mainstreaming contributes to making the approach more effective. You can contribute to the institutional learning by collecting data and information on indicators, reporting on progress and facilitating experience exchange.
- **Gender expertise**: this expertise should be internal, but the use of special external expertise might be considered as well.

Gender mainstreaming is:

- The differences between women and men, girls and boys should never be used as a ground for discrimination.
- A holistic approach towards gender equality that tackles the interconnected causes that generate unequal relationships between women and men, girls and boys in all areas of life.
- Bringing the experience, knowledge, and interests of women and men, girls and boys to bear on the agenda.
- Identifying issues and concerns related to gender equality across all areas of activities.
- Diagnosing the root causes of gender inequality and addressing the differences and disparities through remedial action plans.
- Transforming gender unequal social and institutional structures into equal and just structures for men and women, girls and boys.
- Establishing adequate accountability mechanisms for monitoring progress related to gender equality.

(*Source:* European Commission, 2004)

Gender mainstreaming is not:

- Looking at one gender in isolation.
- About having well written statements – It is about transforming the process of work.
- About women taking action and being the main beneficiaries.
- About targeting activities to support one gender over the other.
- About making specific policies, programmes or projects that are targeted towards women, to be replaced or become obsolete.
- An isolated exercise - It is an integral part of the project or policy making process.

Gender mainstreaming covers

- The design of policies, activities and programmes
- Decision-making

- Access to resources
- Procedures and practices
- Methodology
- Implementation
- Monitoring and evaluation of policies, activities and programmes to assess its impact on both women and men, girls and boys whilst identifying areas which need to be improved.

Example

In considering women's and men's concerns as part of the planning, implementation and evaluation of a project you will find that women and men do not have the same needs in relation to time-schedule, the availability of public transportation, or security related issues, etc., because they have different social and private responsibilities and different levels of mobility. The project then should take into consideration the differences and adapt as necessary in order to mainstream gender issues throughout the whole project cycle.

(*Source:* Organisation for Security and Co-operation in Europe, 2006).

Gender mainstreaming importance

Gender mainstreaming is an important strategy that incorporates a number of specific benefits:

- It places individuals of both genders at the heart of policy-making.
- Gender mainstreaming leads to better governance.
- It involves women and men, boy and girls and makes full use of human resources.
- It makes gender equality issues visible in the mainstream of society.
- It takes into account the diversity among women and men, girls and boys.

Benefits of Gender Mainstreaming

Some of the major benefits include:

- Good gender planning makes good economic sense so that women as well as men are active, using 100% of the productive labour force.
- Studies have shown that discrimination based on gender increases poverty, slows economic growth and weakens governance.
- Promoting gender equality also promotes fundamental human rights.

Basic Principles of Gender Mainstreaming

At the ECOSOC meeting in 1997, the following were stated as the principles for mainstreaming a gender perspective in the United Nations system. The same can be modified to suit different governmental and non-governmental organizations.

- Issues across all areas of activity should be defined in such a manner that gender differences can be diagnosed, that is, an assumption of gender-neutrality should not be made.
- Responsibility for translating gender mainstreaming into practice is system-wide and rests at the highest levels. Accountability for outcomes needs to be monitored constantly.
- Gender mainstreaming also requires that every effort be made to broaden women's participation at all levels of decision-making.
- Gender mainstreaming must be institutionalized through concrete steps, mechanisms and processes in all parts of the United Nations system.
- Gender mainstreaming does not replace the need for targeted, women-specific policies and programmes or positive legislation, nor does it substitute for gender units or focal points.
- Clear political will and the allocation of adequate and, if need be, additional human and financial resources for gender mainstreaming from all available funding sources are important for the successful translation of the concept into practice.

Gender Mainstreaming in Practice

When we want to apply the Gender Mainstreaming strategy, the following three aspects need to be kept in mind:

- **Technical dimension:** Mainstreaming requires reliable data (sex disaggregated data, large sample size, etc.), sound theoretical understanding, gender experts with the ability to spot opportunities and interpret gender equality requirements to varied groups.
- **Political dimension:** Promotion of women as decision makers and supporting women's collective efforts to redefine development agendas; in other words, not increased participation in an unjust and flawed development process, but creating an enabling environment for women that allows them to assess their situations and express their priorities and concerns.
- **Cultural dimension:** It is important to realize that no initiative is operating in a cultural vacuum. The interpretation and implementation of gender mainstreaming initiatives are significantly moulded by socio-cultural practices, and prevailing ideas and ideologies at the local level. These can greatly impact the effectiveness of any project and must be taken into consideration when trying to implement the mainstreaming strategy.

Gender Mainstreaming in Policy Analysis

A gender perspective needs to be included at several points in the policy process:

- In the **formulation of the policy issue/question** to be addressed. If the issue is narrowly defined, the potential for considering gender issues may be reduced. For example, completely different discussions will result in a trade policy discussion focuses only on commodities, rather than examining who produces these commodities and how they are affected by policy choices.
- In the **definition of the information needs** to assess policy option. For example, how can information on both women's and men's situations be incorporated in the decision-making process? There may be a need to ask different kinds of questions and look for information that helps to reformulate or refocus the policy discussion.

- In the **assessment of the implications** of different options by gender. Various options could have different costs and benefits for women and men and different consequences for gender relations and gender equality.
- In the **determination of who will be consulted and how** on matters such as the formulation of the issue, the definition of information needs, and assessment of options. The involvement of civil society, public consultation, polling etc. can have different outcomes depending on who is consulted, what they are asked and how they are consulted or involved. It is important to seek meaningful inputs from both women and men.
- In the **formulation of recommendations** for policy choices. These should reflect the information and analyses on gender equality issues resulting from the previous steps.

Approaches of gender mainstreaming

- **WID Approach**- Improving the material condition of women
- **GAD Approach**- Challenging the power relations and roles within the society
- **Equity Approach**- Focused on women's strategic gender needs
- **Efficiency Approach**- Attempted to harness women's labour for effective development
- **Empowerment Approach**- Focused on women's need those are identified by the women themselves
- **Integration Approach**- Aims at integrating women's awareness and efficiency into mainstream development.

GENDER MAINSTREAMING: AS A STRATEGY

The key elements of any gender mainstreaming strategy are as follows:

- **One** is the necessity to give attention to gender equality from the **initial** stages of processes so that there is scope to influence goals, strategies and resource allocations and thus bring about changes in policies, programmes and other activities and make a real difference to gender equality.

- **Secondly**, the Platform for Action (Beijing 1995) made it very clear that **gender analysis** is the first essential step in the mainstreaming strategy. Before any decisions are taken in any area of societal development, an analysis should be made of the current responsibilities and contributions of women and men and the potential impact of planned processes and activities on women and men respectively.
- **Third**, it is important to note that gender mainstreaming requires a focus on both **women and men**. It is not possible to adequately implement gender mainstreaming by focusing exclusively on women.
- **Fourth** critical factor is that gender mainstreaming requires **explicit** attention to gender perspectives. Gender mainstreaming can never make gender perspectives invisible; in fact this strategy should make them very visible and show the links between gender perspectives and achievement of the overall goals of the policy or programme.
- **Fifth,** gender mainstreaming is not about adding a women's component to existing programmes, and it involves much more than increasing the numbers of women participating. It should **situate gender equality issues at the centre** of policy decisions and resource allocations.
- **Finally**, building on these factors, gender mainstreaming should involve a **transformative** process, which means that incorporation of gender perspectives would identify the need for changes in goals, policies, strategies and actions, as well as institutional changes - changes in structures, procedures and cultures. Once the perspectives of all stakeholders are fully incorporated through gender mainstreaming processes, it is rarely possible to do "business as usual".

Gender Mainstreaming: From Policy to Practice

In order to make gender mainstreaming an important element of any development strategy, the following steps are required:

- The first step required is *an assessment of the linkages between gender equality and the issue or sector*, that is, to identify the gender implications of working on, for example, poverty elimination, agricultural development, energy programmes, peace and security, or enterprise development. This involves understanding why promotion of gender equality is important for achievement of all other development goals.

- Secondly the ***opportunities for introducing gender perspectives need to be identified*** in the work-tasks undertaken. These opportunities or entry-points can be found in research and analysis, policy development, use of statistics, training events, as well as in planning, implementing and monitoring projects and programmes.
- Thirdly an ***approach or methodology*** has to be identified for successfully incorporating gender perspectives into these work-tasks in a manner, which facilitates influencing goals, strategies, resource allocation and concrete outcomes. Different work tasks require different approaches and methodologies.
- ***Institutional development***, in terms of developing guidelines, utilizing gender specialists, providing competence development for all personnel, is also required to support gender mainstreaming.

Methods and Techniques

A multitude of methods and techniques can be used to put gender mainstreaming into practice. This section briefly describes some of these techniques, which are of a general nature and can be applied to a wide variety of different contexts:

- A **baseline gender analysis** assesses the situation at the beginning of an intervention regarding the balance between women and men concerning power, control over resources and access to opportunities.
- A **gender impact assessment** is mostly carried out as part of the preparatory work for a policy or intervention. It assesses the planned undertaking in terms of the different positive and negative effects it has on women, men and gender equality. Questions to be asked in this regard relate to the differences between men's and women's living conditions, human rights, access to justice, political participation, employment, economic situation, security, etc. and how the planned undertaking should take those differences into account.

- In a **gender budgeting** exercise, actors introduce a gender perspective to all levels of the budgetary process and allocate resources to areas that are of specific concern for women to promote gender equality.
- The collection and usage of **gender-disaggregated data** for any decision-making process allows for a better understanding of the decision's implications on women and men. In technical assistance interventions, gender-disaggregated data should be used throughout the project cycle, including for planning, monitoring, reporting and evaluation.
- **Gender-sensitive communication**, including through teaching materials, media, written publications, capacity building measures, etc. can help to avoid perpetuating stereotypes, promote a positive public image of both genders and raise awareness about discriminatory practices.
- Ensuring a **gender-balanced participation** in teams, working groups, decision-making bodies, capacity building measures, public meetings, peace processes, etc. contributes to equal representation, the equal consideration of the perspectives of women and men, as well as the advancement of both genders. Sometimes an adaptation of the arrangements related to an event (e.g. timing, location, procedures, seating layout, etc.) is required in order to achieve a meaningful participation of women.
- The utilization of a **gender-balanced pool of experts and speakers** contributes to an equal visibility and a positive public image of both genders.

Strategies for Mainstreaming Gender with respect to Agricultural Extension

1. Organizing Women Groups.
2. Male extension workers can be trained to work more closely with women in settings that are culturally acceptable, such as women groups. Such groups can also improve access to infrastructure.
3. Technologies to reduce energy and time spent, particularly the household and farm production activities. Extending the technological innovations, such as weeders, paddy threshers, winnowers, sprayers, harvesting tools, parboiling units, maize

shellers, dal making machines, etc., will reduce the burden of women.

4. Increasing the bio-mass production to meet fuel needs, planting of fast growing fodder in common lands and developing mechanisms for sharing the fodder helps women in saving lot of time and devote this time for income generating activities.
5. Innovative credit programmes using non-traditional forms of collateral and local institutions (women groups) can ensure that women are able to obtain access to credit.
6. Identifying the right training and extension needs of women is one of the most important steps in initiating any developmental programme.

   Gender Analysis: Gender Analysis of activities, resources, constraints, implications and benefits should be understood using Participatory Approach. This information should be taken into consideration for needs assessment. The ability of staff members / extension functionaries to do this has to be built up.
7. Giving women farmers more access to meetings, trainings, exposure visits and demonstrations and organizing training programmes based on the needs of the women. Institutional and village based trainings to be organized as per the convenience of the women farmers.
8. Where severe fragmentation exists, collective farming should be encouraged by women.
9. Farmer to farmer training or participatory training should be encouraged.
10. Active women can be selected, trained and provided with inputs and credit to practice improved technologies. Their fields can be used as demonstration plots for training other women.
11. Recruiting more women extension workers from rural areas and training them.
12. Female para-extension agriculturists, who are relatively uneducated, can be posted in their own villages with short crash courses on agriculture.
13. Appointing female supervisors and Subject-matter specialists

(SMSs).

14. Making better use of male extension agents.
15. Change the stereo-typed attitude of male agents with regular gender sensitization courses.
16. Increasing awareness of gender roles
17. Developing skills in use of language and communication to suit women
18. Using women as contact farmers or farmer friends.
19. Crèches for children of women farmers. This will enable girls to go to schools.
20. Proper health care support for girls and women.
21. Most of the micro enterprises undertaken by women are based on the skills and raw material available rather than considering the market needs and market dynamics. In-depth marketing study would help identify effective marketing strategy for products. Co-operative marketing of products and assigning brand names for the products would also be helpful in finding sustainable markets.

❑❑❑

Chapter 9

Gender Sensitization

Gender Sensitization is one basic requirement for the normal development of an individual. Without being sensitive to the needs of a particular gender, an individual may refrain from understanding the opposite gender and in some acute cases even him or herself. The need for this sensitivity has been felt and realised through times immemorial and in almost all kinds of human existence, across the globe.

Gender sensitization refers to the modification of behaviour by raising awareness of gender equality concerns. Gender sensitization theories claim that modification of the behaviour of teachers and parents towards children can have a causal effect on gender equality. Gender sensitizing "is about changing behaviour and instilling empathy into the views that we hold about our own and the other gender. It helps people in "examining their personal attitudes and beliefs and questioning the 'realities' they thought they know."

Gender sensitisation is a weapon towards Gender Equity, Equal Opportunity and Women Empowerment which is essential for creating sustainable development as enshrined in our Constitution. The State & Central Government are making continuous efforts to realize these promises through Policies & Programs. The progress achieved in this direction has a mixed outcome and there is still long way to realize the objectives of the Equitable Society, Gender Equity and inclusive Development without discrimination which can be achieved bringing awareness among the people about the Rights given under the Constitution and the pro- women Acts and laws affecting women's right and empowerment in India involves a complex pattern of advances and setbacks.

The pitfalls in development due to Gender inequalities, discrimination, denial/deprivation/in accessibility/violence against women are now cannot be overlooked or sidelined. The nations are well aware of the situation arising out of Gender inequalities and discrimination. The Civil Societies are making efforts by pressing the Governments in power to address these issues. But the magnitude of the problem is so high we cannot wait the situation to take its course as the danger due to neglect issues related to Gender is so phenomenal and colossal.

The pro-active initiation towards Gender sensitisation is to ensure adoption of Gender Budgeting at all levels in the letter & spirit. Participation of Women in the processes of Governance is very strongly felt to fulfill the above aspirations. Commitment from the state to ensure women participation in Governance at all levels should be seen in action than in paper.

The attitudinal problem of the society against women due to socio-cultural system detrimental to the safety and security of women are posing great threats. Women are pushed to be mute spectators in all the above situations living them in a hopeless situation. It is time to stop shedding tears for all the evils forced on women and get in to more civilized and responsible acts leading women to a life with dignity, equity, opportunity and empowerment so that they play major role in building a strong nation. Gender equality is, first and foremost, a human right. A woman is entitled to live in dignity and in freedom from want and from fear. Empowering women is also an indispensable tool for advancing development and reducing poverty. Empowered women contribute to the health and productivity of whole families and communities and to improved prospects for the next generation.

The Constitution provides for Gender Equity, Opportunity and Empowerment by creating special provisions to overcome the gaps. The success is patchy and still the large percentages of women are suffering poverty, poor health, oppression and violence. Governance cannot ignore this alarming situation. It is time to re-look our Policies, Programs and the societal structure.

Key issues and linkages:

- Reproductive health
- Stewardship of natural resources
- Economic empowerment

- Educational Empowerment
- Political Empowerment
- Women's Work and Economic Empowerment
- Inter-generational gender gaps
- Empowering Women through Education
- Legal Issues like Domestic Violence Act, 2005, Sexual Harassment (Prevention) Act, 2013
- Police Response to Women's Issues
- Women Issues dealt by Supreme Court of India, such as:
  - Delhi Domestic Working Women Forum Vs. Union of India and Others JT 1994 (7) SC 183
  - Government of Andhra Pradesh Vs. P.B. Vijay Kumar and Another Air (1995) SC 1648
  - Sarla Mudgal Vs. Union of India (1995) 3 SCC 635
  - Raghubar Singh and Others Vs. Gulab Singh and Others Air (1996) SC 2401
  - Vishaka and Others Vs. State of Rajasthan and Others Air 1997 SC 3011
  - Gaurav Jain Vs. Unioin of India Air (1997) SC 3021
  - Githa Hariharan and ANR Vs. Reserve Bank of India and AnR Air (1999) 2 SCC 228
  - Apparel Export Promotion Council Vs. A.K.Chopra Air 1999 SC 625
  - Municipal Corporation of Delhi Vs. Female Workers (Muster Roll) Air 2000 SC 1274
  - Buddhadev Karmaskar Vs. State of West Bengal (Date of Judgement - 15.9.2011)
- Awareness on **Legal Rights Women,** such as Free legal aid, Right to privacy while recording statement, No arrests after sunset, Protect your identity & Employers must protect.

Gendered assumptions and gender-based social discriminations exist throughout society, from organisational planning to the everyday interactions at the workplace. Negative impacts of such assumptions can undermine the efficiency and success of even the best-intentioned plans.

Gender sensitization training is the practical tool for analysing gender relations and provides adequate information regarding major factors that influence and are responsible for maintaining or change the structure of gender relations. It is also a development intervention which aims at awareness, knowledge, skill and behaviour in relation to gender. It concerns with developing skills, capacity to translate such awareness into very specific tools that can be used into practice.

Gender sensitization will help in:

- Reviewing existing policies, rules, regulations and procedures with a view to encompassing gender values, needs and aspirations
- To formulate Gender Perspective Guidelines
- To plan, implement, monitor and evaluate gender specific initiatives
- To co-ordinate and monitor gender activities in all ministries

Why Gender Sensitization?

- Gender sensitivity helps to generate respect for the individual regardless of sex.
- Gender sensitivity is not about pitting women against men. On the contrary , education that is gender sensitive benefits members of both sexes. It helps them determine which assumptions in matters of gender are valid and which are stereotyped generalizations.
- Gender awareness requires not only intellectual effort but also sensitivity and open mindedness. It opens up the widest possible range of life options for both women and men.

Why is Gender Sensitivity Training at Workplace is Important?

- The contemporary workplace is very diverse and is becoming more so every day. Employees need to understand, be sensitive, and be able to adapt to the various needs, concerns, and characteristics of a multitude of different people including women.
- Gender Sensitivity training will help employees to personally cultivate good interpersonal relationships with members of their

team and help facilitate respective and productive group relations among team members.

- It helps to develop a cordial and amiable working environment where there is mutual respect and trust between the sexes, without sexual innuendoes and obstructive chauvinism to mar the atmosphere.

Objective Gender Sensitivity Training

- The primary objective of gender sensitivity training is to educate the participants about the needs and concerns of women for more constructive behavior that will be beneficial to you and everyone else in the organization.
- It helps an individual by providing insight into your behaviour and helps you develop corrective emotional and behavioral actions.

Importance of Gender sensitization

- Gender sensitivity helps to generate respect for individual irrespective of sex.
- It also indicates the many challenges that remain in order to close gender gaps, promote equality in education, employment and other spheres of work and family.
- According to modern trends, men and women are equally active players in the family, community, and national affairs. Women must be recognized as representing half of the world's most precious resources called human beings.
- Employers should commit to equal employment opportunity at workplace and management should not distinguish people on the on the basis of sex. For all management purposes, both men and women are gifted with the ability to work with diligence.
- Managements must sensitize employees for gender sensitivity to prevent or minimize issues connected with sexual harassment that is illegal and train employees to be more vocal and how to say no to such advances firmly.

❑❑❑

Chapter 10

Gender Budgeting

A gender budget is **not** a separate budget for women. Instead, the gender budgets are an attempt to assess government priorities as they are reflected through the budget and examine how they impact women and men.

Gender budgets look at what the impact of spending is on men and women and whether or not budgets respond to the needs of both women and men adequately.

"Women's budgets", "gender budgets", "gender-sensitive budgets", and "gender responsive budgets" are all terms that are used to describe initiatives that have used gender as lens from which to analyse budgets at national, regional, and civic levels.

Need of a Gender Budget

Gender Budget Initiatives are attempts to disaggregate the government's mainstream budget according to its impacts on women and men. It refers to the process of conceiving, planning, approving, executing, monitoring, analysing and auditing budgets in a gender-sensitive way. The gender budgeting exercise would potentially assist and lead to the following empowering measures:

- Addressing gap between policy commitment and allocation for women by emphasizing on adequate resource allocation.
- Putting pressure and focus on gender sensitive programme formulation and implementation.
- Mainstreaming gender concerns in public expenditure and policy.
- By being a tool for effective policy implementation where one can check if the allocations are in line with slated gender sensitive policy commitments and are having the desired impact.

Gender budget is helpful in:

- Improving women's economic equality.
- Improving effectiveness, efficiency, accountability, and transparency of government budgets.
- Revealing discrepancies between what a governments says it is doing and the actual impact of government policies.
- Offering a practical way for the governments to implement their obligations under international human rights agreements, such as the Convention on the Elimination of All Forms of Discrimination against Women (CEDAW).

The Scope of Gender Budgeting

Gender Budgeting expands our concept of the economy to include things that are not usually valued in money. In particular, Gender Budgeting recognizes the unpaid care economy - the work that mainly women do in bearing, rearing and caring for their families and the people in our society. Gender Budgeting recognises that unless this unpaid care work is done, the economy will not function effectively and people's well-being will be very negatively affected. Government therefore needs to find ways of supporting those who do this unpaid care work, lessening their burden, and ensuring that the work is done well. Gender Budgeting should, however, not be confined to the 'social' or 'soft' areas such as education, health and welfare. Gender Budgeting is a tool for gender mainstreaming in the developmental process as a whole. As such, it needs also to be applied in areas such as agriculture, power, defence, commerce, and information technology where the gender implications may not be immediately apparent. Later chapters give examples of how this can be done.

Gender Budgeting Around the World

- The concept of gender budgeting is a nineties' trend that has been introduced *mostly in Commonwealth countries.*
- Australia was the first country to implement a women's budget in 1984. Federal, state and territorial governments in Australia examined the impact of budgets on women and girls for 12 years until a change of government in 1996.

- South Africa's Women's Budget Initiative was initiated in 1995 and involves NGOs, parliamentarians, and a wide range of researchers and advisors. Gender budget initiatives in Tanzania (1997) and Uganda (1999) examine the impacts of structural adjustment programs in these countries and specifically focus on education and health.
- Many of the earlier gender budget initiatives focused primarily on the expenditure side rather than the revenue side of government budgets. Since 1995 there have been gender budget initiatives in more than 60 countries around the world.

How Gender Budgeting helped the Governments around the world?

- In Philippines, they made a specific requirement that every government agency allocate at least five per cent of its budget to gender and development initiatives.
- UK '*From the wallet to the purse*'. In United Kingdom, the government announced that from 2003 onwards the new Child Tax Credit would be paid to the main career – usually a woman – rather than to the main earner – usually a man. The group supporting this used the slogan "From the wallet to the purse (men carry wallets while women carry purses) to argue that giving money to women was more efficient and in-line with government policy on reducing child poverty.
- In South Korea, a gender budget initiative demonstrated that most of the beneficiaries of training and education programs were leaders or women from women's organizations.

Gender Budgeting Statement (GBS)

The Gender Budgeting Statement (GBS) which *comprises the gender specific demands for grants,* has emerged as an important advocacy tool which reflects on the flow of funds for women and encourages debate and discussions on Gender Budgeting.

Gender Responsive Budgets

A gender responsive budget is a budget that acknowledges the gender patterns in society and allocates money to implement policies and programmes that will change these patterns in a way that moves towards a more gender equal society. Gender budget initiatives are exercises that aim to move the country in the direction of a gender responsive budget.

Gender Budgeting in India

In 2004, the Ministry of Women and Child Development (MWCD) recognized Gender Budgeting as an important tool for women's empowerment, and as a way of addressing the observed inequality. The MWCD adopted "Budgeting for Gender Equity" as a mission statement, and framed a Strategic Framework of Activities to implement this mission.

The Ministry of Finance, as the agency with primary responsibility for the budget, has also recognized the potential of Gender Budgeting. The Ministry of Finance mandated all ministries to establish Gender Budgeting Cells by January, 2005, and asked 18 ministries and departments to submit a report highlighting budgetary allocations for women. The first Gender Budget Statement (GBS) appeared in the Union Budget, 2005-06.

Gender Budget Cells

The Department of Expenditure, Ministry of Finance issued a Charter for the Gender Budget Cells, on 8 March, 2007, clearly articulating the composition and the functions of the gender budget cells. The MWCD has been actively pursuing other Ministries with regard to setting up of gender budget cells. These gender budget cells serve as focal points for coordinating Gender Budgeting initiatives both intra- and inter-ministerially. The roles envisaged for these cells include:

- Act as a nodal agency for all Gender Budgeting initiatives.
- Pilot action on gender sensitive review of public expenditure and policies (Expenditure/Revenue/ Policies/Legislation etc.)
- Guide and undertake collection of sex disaggregated data for target group of beneficiaries covered under expenditure, revenue raising/ policy/ legislation.
- Guide Gender Budgeting initiatives within departments as well as in field units responsible for implementing government programmes.

- Conduct gender based impact analysis, beneficiary needs assessment and beneficiary incidence analysis to:
  - Establish effectiveness of public expenditure
  - Identify scope for re-prioritization of public expenditure
  - Improve implementation, etc.
  - Collate and promote best practices on participative budgeting for and implementation of schemes

Data Needs for Gender Budgeting

Gender Budgeting relies heavily on data, so that policies, programmes and budgets can be evidence based rather than based on myths or assumptions. Data are needed at different stages of the budgeting process:

1. Initially data are needed to describe the situation of women and men, girls and boys.
2. Subsequently, data are needed to reflect delivery of the programmes and projects i.e., how many women and men, girls and boys are reached.
3. Data are also needed to show the impact of the policies and programmes i.e., whether they have made a difference to the situation of the people of the country. Thus, sex disaggregated data is a basic requirement for GB which can be collected using different gender analysis tools.

Five Steps Framework for Gender Budgeting

Step 1	An analysis of the situation for women and men and girls and boys (and the different sub groups) in a given sector.
Step 2	An assessment of the extent to which the sector's policy addresses the gender issues and gaps described in the first step. This step should include an assessment of the relevant legislation, policies, programmes and schemes. It includes an analysis of both the written policy as well as the implicit policy reflected in government activities. It should examine the extent to which the above meet the socio-economic and other rights of women.
Step 3	An assessment of the adequacy of budget allocations to implement. The gender sensitive policies and programmes identified in step 2 above.

Step 4	Monitoring whether the money was spent as planned, what was delivered and to whom. This involves checking both financial performance and the physical deliverables (disaggregated by sex)
Step 5	An assessment of the impact of the policy / programme / scheme and the extent to which the Situation described in step 1 has been changed, in the direction of greater gender equality.

(*Source:* UNIFEM-UNFPA Gender Responsive Budgeting and Women's Reproductive Rights: Resource pack)

Based on the above steps there are number of Entry Point Tools that can be used to do Gender Budgeting. Some of the most appropriate in the present Indian context are mentioned below. You can use either one or a combination of two or more of these tools, at any given time.

1. Participatory planning and budgeting

- Involvement of women in decision making is seen as a key to the success of Gender Budgeting initiatives.
- It is not enough to say that women's needs and concerns have been taken into account.
- Women must be treated as equal partners in decision making and implementation rather than only as beneficiaries. For this there may be need for capacity building.

2. Spatial mapping

Women's empowerment requires adequate resource allocation in all areas including health, education, water sanitation and nutrition, sustained employment, access to credit and asset ownership, skills, research and design technology and political participation. Regional imbalances (including rural-urban) must be corrected. Spatial mapping will help in assessing resources required taking into account size of population and norms for availability of facilities, so that universalisation of basic socio-economic infrastructure is achieved progressively, and allocations and interventions are more focused.

3. Gender appraisal for all new programmes and schemes

All new programmes, projects and schemes (PPS) should be passed through a gender lens. This will ensure that gender sensitivity and women's participation is built into new all new PPSs from the start. It will also ensure that the implementation modality of the PPS is gender sensitive and that gender impact assessment is built into the design of the PPS.

4. Guidelines for gender-sensitive review of public expenditure and policy

The Ministry of Women and Child Development has formulated guidelines in the form of Checklists I and II. Checklist I is for programmes that are beneficiary oriented and consciously target women. Checklist II covers other 'mainstream' sectors. These guidelines help in reviewing public expenditure and policy from a gender perspective to enable identification of constraints in the reach of programmes and policies and inform suitable corrective action.

5. Gender-based profile of public expenditure

Preparation of the profile facilitates review of schemes and other public expenditure from a gender perspective and identifies gender component by way of expenditure and physical targets. The trend of the gender component is indicative of extent to which budgeting is gender responsive. Preparation of the profile will highlight constraints, such as non-availability of sex-disaggregated data as well as the need to determine reasons for non-adherence to targeted expenditure on women.

6. Impact analysis

It is essential to evaluate the impact of programmes and schemes of the government and to see if they are meeting the objectives for which they were started. Results of impact assessment of programmes from a gender perspective can help to identify barriers for women in accessing public services/expenditure.

7. Revenue incidence analysis

This tool shows the proportion of income paid in taxes and user charges by different categories of individuals.

❑❑❑

Chapter 11

Gender Budgeting and India's International Commitments

During the last decade, a number of international meetings have been convened that have the potential for transforming the reality of women's lives. At many of these meetings government, including Government of India, have committed to taking action to improve the situation of women. The following are among the international commitments to which Government of India is party:

- The Convention on the Elimination of All Forms of Discrimination Against Women (CEDAW), which Government of India signed in 1980.
- The World Conference on Human Rights in Vienna (1993) asserted that women's rights are human rights.
- The International Conference on Population and Development (ICPD) in Cairo (1994) placed women's rights and health at the centre of population and development strategies.
- At the Fourth World Conference of Women in Beijing (1995), governments declared their determination "to advance the goals of equality, development and peace for all women everywhere in the interest of all humanity".
- In 1995 Common wealth Plan of Action on Gender and Development governments declared their vision of a world "in which women and men have equal rights and opportunities in all stages of their lives".

"Budgeting for Gender Equity": Strategic Framework of Activities

- Quantification of allocation of resources for women in the Union, States and Local Administration budgets expenditure thereof.

- Refining and standardising methodology and development of tools.
  - Trend analysis
  - Analysis of change in pattern, shift in priorities in allocation across clusters of services, etc.
  - Variations in allocation of resources and actual expenditure
  - Adherence to physical targets
  - Gender audit of policies of the Government - monetary, fiscal, trade, etc. - at the Centre and State levels
  - Research and micro studies to guide macro policies like credit policy, taxes, etc.
  - Identification of gender impact of policies/interventions viewed as gender neutral
  - Micro studies to identify need for affirmative action in favour of women towards correcting gender imbalances
  - Impact assessment of various schemes in the Union and State budgets
  - Micro studies on incidence of benefits
  - Analysis of cost of delivery of services
- Analyzing programmes, strategies, interventions and policy initiatives from the perspective of their impact on status of women as reflected in important macro indicators like literacy, maternal mortality rate (MMR), participation in workforce
  - e.g. analysis of substance and content of various interventions directed at health of women and correlate the same with indicators like MMR to establish need for corrective action in formulation of scheme/approach.
- Institutionalizing the generation and collection of gender disaggregated data:
  - Developing management information system (MIS) for feedback from implementing agencies.
  - Inclusion of new parameters in data collection in Census and surveys by National Statistics Organisation (NSO), Census Statistics Organisation (CSO), etc.

- Consultations and capacity building
  - Collation of research and exchange of best practices
  - Developing methodologies and tools for dissemination
  - Forums and partnerships amongst experts and stakeholders.
- Review of decision-making processes to establish gender equity in participation - review of extent of participation of women in decision-making processes and to establish processes and models aimed at gender equity in decision making and greater participation of women.
  - Formulation and reflection of satellite accounts to capture the contribution of women to the economy by way of their activities in areas that go unreported like care economy, unpaid work in rearing domestic animals, etc.

An Example of Gender Budgeting in Chiranjeevi Yojana

Using the five steps of Gender Budgeting to assess Chiranjeevi Yojana, Gujarat

This example using the Chiranjeevi Yojana - an innovative scheme of the Government of Gujarat to reduce the maternal mortality rate in the state - shows how we can use the five steps of Gender Budgeting to assess a programme for its gender sensitivity:

1. Describe the situation of women and men, girls and boys, who are served by a particular sector.

   The Government of Gujarat was very concerned about the high levels of maternal mortality and infant mortality in the state, with a maternal mortality rate (MMR) of 378 per one lakh live births, and an infant mortality rate of (IMR) of 38 per one lakh live births. To identify the extent of the problem, a needs assessment was undertaken in the five worst affected districts of Gujarat. The assessment revealed that many of the districts did not have adequate infrastructure or skilled doctors to address the issue.

2. Examine government policies and programmes in the sector, to see whether they address the 'gender gaps' - that is, inequalities in the service offered to each group as described in the first step.

The state realised that it had neither a specific programme nor the capacity in terms of infrastructure to address the issue of MMR completely on its own. The State Government of Gujarat, in its Vision-2010 document, indicated that it would reduce maternal mortality from 378 to 100 per one lakh live births and reduce the IMR from 38 to less than 30. Hence, they formulated Chiranjeevi Yojana to provide facilities for institutional deliveries of women in below poverty line (BPL) families using the private sector. The Health Department identified private institutions and doctors who could provide these facilities and the State signed a memorandum of understanding with the private practitioners to do what was needed. Both electronic and print media were used to publicise the scheme amongst the target group.

3. Examine the budget to see whether sufficient money has been allocated to implement effectively the gender sensitive policies and programmes.

   The scheme was started in the five worst affected districts of Kutch, Banaskanta, Sabarkantha, Godhra and Dahod in December 2005. Rs. 93.5 million was allocated in the first year to provide delivery services to an estimated 58,637 pregnant women. The average cost of each delivery and transportation cost worked out to Rs 1,795 (assuming there would be about 20 complicated deliveries, for which a higher rate was fixed, for every 80 normal deliveries).

4. Monitor whether the allocated money has been spent and who has benefited from the money.

5. Go back to the first step and re-examine the situation, to see whether the budget and its associated programme have improved on what was initially described.

   The Ministry of Health undertook a review within six months of the implementation of Chiranjivi Yojana to evaluate its effectiveness through a rapid qualitative study. The specific objectives were to examine whether there was increased accessibility of services; understand the utilisation pattern of services, client satisfaction, analyse the factors influencing and impeding service utilisation. Views of programme managers were gathered to suggest effective ways of improving uptake of services and to explore the possibilities of expanding the package of services under the scheme. It was found that during the first six months of the implementation of the Scheme, only one maternal death was reported among the beneficiaries who benefited from the Scheme. There were, however, some areas for improvement:

- It was suggested that there was need to expand the package of services and include care for the new-born who often required medical attention but were not covered under the scheme.
- The MOU between government and private providers discussed partnership and payment but did not explicitly state details of elements covered in the CY package. This led to problems regarding reimbursement of persons accompanying the women to hospital.
- There were no guidelines on the use of standard definitions for obstetric complications and quality compliance.
- Change in modalities of payment from advance payment to reimbursement to private providers led to problems of payment.
- Revisions were made based on the review as well as feedback from beneficiaries (existing and potential) and service provider. The Scheme is now universally implemented across the State.

(*Source:* Benita Sharma, IFES-USAID, based on the evaluation of the Chiranjeevi Yojana by UNFPA)

Case Study: Gender Budgeting In Agriculture

(This case study is based on research conducted for the Gender Resource Centre of the Ministry of Agriculture, GOI by Neeraj Suneja)

Rural women are major producers of food in terms of value, volume and hours of work. Nevertheless, women's control over resources and access remains extremely limited. Women may function as head of the household for the major part of the year. Nevertheless, the landlords and officials continue to recognise the husband as the cultivator in the official lists and statistics. Women then have difficulty in accessing credit and inputs from mainstream institutions and government schemes and also in becoming members of farmers associations and beneficiary organisations. Agricultural research has also focused on increasing the production of high value major cereal and cash crops rather than the traditional varieties of cereals and subsistence crops which are farmed by women and which provide the major source of food to their families. Upgrading of technology has focused on implements and tools designed with male users in mind.

The National Agriculture Policy of 2000 gave high priority to 'recognition and mainstreaming of women's role in agriculture'. At state level, states are encouraged to allocate 30% of allocations for women farmers and women extension functionaries under the extension interventions, focusing on formation of Women SHGs; capacity building interventions; linking women to micro credit; and improving their access to information through IT and other extension activities. At the central level, a National Gender Resource Centre in Agriculture (NGRCA) has been established to assist the centre and the states with advisory services.

The Ministry of Agriculture has started a number of programmes and schemes which target women. These include:

Horticulture: The State Horticulture Missions have been directed to earmark at least 30% of their budgets for women beneficiaries in all ongoing programmes under the National Horticulture Mission and Technology Mission for Horticulture in North Eastern States, Sikkim, Jammu & Kashmir, Himachal Pradesh and Uttaranchal.

Agriculture Extension: In the scheme "Support to States Extension Programme for Extension Reforms", 30% of resources are meant to be allocated for women farmers and extension functionaries.

Watershed Development Programmes: The Watershed Development programmes provided the involvement of women farmers in the constitution of Watershed Associations, other institutional arrangements, formation of women SHGs and User Groups (UGs).

Crops: The scheme "Technology Mission on Cotton" encourages states/implementing agencies to give preference to women farmers in components like distribution of agriculture inputs, trainings and demonstrations so that at least 20% of the total allocation reaches them. Under another Mission, a subsidy is provided for the distribution of sprinkler sets to women farmers and other disadvantaged groups.

Technology Mission on Oilseeds & Pulses: The "Integrated Scheme of Pulses, Oilseeds, Palm Oil and Maize" provides subsidy/ assistance to women farmers for sprinkler sets and pipes for carrying water from source to the field.

Integrated Nutrient Management: 25% of seats are reserved for women in the training courses for farmers on organic farming.

Some states have also initiated schemes targeting women. For example, a scheme launched in the Tenth Plan by the Extension Division namely "Support to States for Extension Reforms" provides for representation of women in all bodies at district level, including the governing board, farmer advisory committees, farm women interest groups and commodity-based organisations.

Different players / stakeholders in Gender Budgeting initiatives

There are a range of different actors who can be involved in Gender Budgeting. They have different roles and carry out different activities. Some of them are:

- The Ministry of Women and Child Development(nodal ministry at the central level, in India)
- The Ministry of Finance (at the Centre and in the States)
- The Planning Department or Planning Commission (at the Centre and in the States)
- Sectoral ministries - each and every department and ministry can do Gender Budgeting – albeit some have more possibilities than others (at the Centre and in the States)
- Researchers and economists
- Statisticians
- Civil society organizations especially women's groups
- Parliamentarians, MLAs and other representatives of the people at state/district and sub-district levels.
- Media
- Development partners/donors
- The women and men for whom the specific policy, programme or budget is intended

Difference between Gender Budgeting and Gender Auditing

Gender auditing is part of the Gender Budgeting process. Gender auditing is the process that is conducted after the budget has been implemented. It is the process of reviewing financial outlays - looking at trends over time, percentage shares, etc; analysing and assessing systems actually put in place, processes adopted, outcomes and impacts of budgetary outlays vis-à-vis what was planned - all this through a gender lens.

❑❑❑

Chapter 12

Gender Audit

Gender Audit

Women and men from various social groups are positioned differently in society- in the workplace and family- and have different responsibilities and needs. Due to this, women and men have different experiences, knowledge, talents and needs. Therefore, different programmers/schemes are designed to address this imbalance. Public services like education, health, transportation and welfare are utilized differently by women and men; as a result budgets impacts men and women differently.

UNESCO defines gender audit as management and planning tool. Gender Audit evaluates the gender responsiveness of an organization's culture and how well the organization is integrating the gender perspective into its work. The audit outcome aims to assist the organization to become more gender responsive. It is also a kind of impact evaluation that assesses or measure the impact of interventions on gender equality and women's empowerment. Traditionally audits are commonly associated with accounting audits in the financial world, used to certify that the finances and administration are legitimate, with established rules and regulations correctly followed. In the 1980s, quality management audit was introduced in companies to promote improvement of company performance. Following the same principles, social audits have been developed in a range of organizations to measure the extent to which they live up to "the shared values and objectives" to which they are committed. Social audit is a framework which allows an organization to build and develop a process whereby it can account for its social performance, report on that performance and draw up an action plan to improve on that performance. Gender Audit is one such form of social audit. Therefore, it can be summed up by stating that Gender Audit is a guided process to examine "what has been done" and "what has not been done to meet the Governments stated objectives of gender equality.

Characteristics of Gender Audit

1. Gender Audit takes into account whether internal practices and related support systems for gender mainstreaming are effective and reinforce each other.
2. Monitors and assesses the relative progress made in promoting gender sensitivity
3. Identifies critical gaps and challenges
4. Recommends ways of addressing these challenges and suggests new and more effective strategies to overcome gaps
5. Documents good practices towards the achievement of gender equality

A Gender Audit provides tools to audit the processes which have been undertaken by any Department or organizations under the various policies, schemes and programmes. It examines whether and at what level Gender Budgeting initiatives have been adopted in the planning, implementation and review of the policies and budgetary allocations of the Department. The results of Gender Audit can provide learning and guidelines for Gender Budgeting to revisit policies, programmes and schemes as well as implementing mechanisms to ensure Gender mainstreaming. The Gender Audit will help the institution to identify gaps between their own gender equality policy and practice, and the way their programmes impact on gender equality.

ILO (A manual for gender audit facilitators: The ILO participatory gender audit methodology, 2007) defines audit as a tool to assess the extent of Gender mainstreaming accomplished - it helps to assess the differential impact of projects and policies on women and men. Its aim is to see whether the policy, programme guidelines, practices, systems, procedures and budgets are being used in the most effective way to deliver the Governments commitments to gender equality.

It states that a gender audit:

- Considers whether internal practices and related support systems for gender mainstreaming are effective and reinforce each other and whether they are being followed ;

- Monitors and assesses the relative progress made in gender mainstreaming ;
- Establishes a baseline ;
- Identifies critical gaps and challenges ;
- Recommends ways of addressing them and suggests new and more effective strategies ;
- Documents good practices towards the achievement of gender equality.

Here in a gender audit is essentially a "social audit", and belongs to the category of "quality audits", which distinguishes it from traditional "financial audits". It considers whether internal practices and related support systems for gender mainstreaming are effective and reinforce each other and whether they are being followed. The belief is that a gender audit enhances the collective capacity of the organization to examine its activities from a gender perspective and identify strengths and weaknesses in promoting gender equality issues. It monitors and assesses the relative progress made in gender mainstreaming and helps to build organizational ownership for gender equality initiatives and sharpens organizational learning on gender through a process of:

- Team building
- Information sharing
- Reflection on gender

(*Source:* ILO Participatory Gender Audit)

Difference between Gender Audit, Gender Assessment and Gender Analysis Study

Gender Analysis Study	Gender Assessment	Gender Audit
Conducted during program design or planning	Conducted during project implementation or as part of the final evaluation after a project has been completed	Conducted during project implement-ation. Assesses the internal and institutional context in which the programoperates
Analyzes the gender context in which the	Assesses the program (both technical	Evaluates gender integration as it

program is operating	processes and result)	relates to: policies, staff capacity, tools, training and resources, organiz-ational culture and workplace issues
Helps programmers understand gender roles, responsibilities, statuses and inequalities so they can use that inform -ation to design, implement,monitor and evaluate programs	Assesses the extent to which the program integrates gender issues into activities and institutional systems. Assesses program results and effects on male and female participants Conducted via similar research processes as gender analysis, but includes a comprehe-nsive review of existing activities and initiatives.	Conducted via: desk review, question-naire, interviews and focus groups with staff. Should ideally be carried out by external consultants. Makes recommendations-about how to better mainstream gender within organization and among staff.
Conducted via: desk review, key informant interviews and interviews/focus groups with beneficiaries	Includes recommend-ations based on results and lessons.	
Includes recommend -ations on how to integrate Gender considerations into the program	Can include components of a gender analysis study, particularly if one has not already been done	

A gender audit enhances the collective capacity of the organization to examine its activities from a gender perspective and identify strengths and weaknesses in promoting gender equality issues. It monitors and assesses the relative progress made in gender mainstreaming and helps to build organizational ownership for gender equality initiatives and sharpens organizational learning on gender. To do this, it:

- Considers whether internal practices and related support systems for gender mainstreaming are effective and reinforce each other and whether they are being followed;

- Monitors and assesses the relative progress made in gender mainstreaming;
- Establishes a baseline;
- Identifies critical gaps and challenges;
- Recommends ways of addressing them and -suggests new and more effective strategies;
- Documents good practices towards the achievement of gender equality.

A participatory gender audit is a tool and a process based on a participatory methodology to promote organizational learning on how to practically and effectively mainstream gender in policies, programmes and structures and assess the extent to which policies have been institutionalized at the level of the:

- Organization
- Work unit
- Individual

The participatory gender audits' objectives are to:

- Generate understanding of the extent to which gender mainstreaming has been internalized and acted upon by staff;
- Assess the extent of gender mainstreaming in terms of the development and delivery of gender-sensitive products and services;
- Identify and share information on mechanisms, practices and attitudes that have made a positive contribution to mainstreaming gender in an organization;
- Assess the level of resources allocated and spent on gender mainstreaming and gender activities;
- Examine the extent to which human resources policies are gender-sensitive;
- Examine the staff sex balance at different levels of an organization;
- Set up the initial baseline of performance on gender mainstreaming in an organization with a view to introducing an ongoing process of benchmarking to measure progress in promoting gender equality.

Chapter 13

Self-Help Group

"Small, economically homogeneous affinity groups of rural poor voluntarily formed to save and mutually contribute to a common fund to be lent to its members as per the group members decision." (NABARD)

Self-Help was coined by Samuel Smiles

SHG - gained significance by Prof. Mohammad Yunus

The concept of self-help groups had its origin in the co-operative philosophy and the co-operators by and large, including the National Federations in the credit sector, could not think of any better SHG than a primary co-operative credit society itself. As SHG are small and economically homogenous affinity groups of rural poor, they are voluntarily coming together for achieving the following:

1. To save small amount of money regularly.
2. To mutually agree to contribute a common fund.
3. To meet their emergency needs.
4. To have collective decision-making.
5. To solve conflicts through collective leadership mutual discussion.
6. To provide collateral free loan with terms decided by the group at the market driven rates.

A self-help group is also defined as a *"self-governed, peer controlled information group of people with similar socio-economic background and having a desire to collectively perform common purpose."*

Historical Evolution of SHGs

- By the late 1980's, Mysore Resettlement and Development Agency (MYRADA), a NGO, located in South India, formed around 300 autonomous SHGs known as the Credit Management Groups.
- The National Bank for Agricultural and Rural development (NABARD), found SHG as an interesting and effective strategy to provide banking services to the so called 'un-bankable' people.
- Between 1991 and 1992, NABARD in consultation with RBI, Commercial Banks (CBs) and NGOs launched the pilot project of linking the SHGs with Commercial Banks, based on NABARD guidelines.
- By the late 1990s and early 2000, given its success the government had become a key promoter of SHGs. Decentralization of power at the Panchayt's level in 2004 gave local bodies more teeth and SHGs came to be recognized as a powerful institution for the poor.
- RBI had advised Commercial Banks in July 1991 to extend finance to SHGs as per NABARD guidelines. Subsequently, the linkage project was extended to RRBs and Co-operatives.

The distinguishing features of self-help groups are:

a) An SHG normally consists of not less than five persons (with a maximum of twenty) of similar economic outlook and social status.

b) It promotes objectives like economic improvement and raising resources for development and freedom from exploitation.

c) It has its own by-laws for the proper functioning of the group as well as for the observance of certain rules by the group members and regulations concerning membership.

d) The form of such a group could be mostly on an informal basis (unregistered).

e) Periodical meetings of members are held for solving their problems (economic and social) and they collect fixed savings of the members.

f) The savings of members are kept with a bank in the name of group and authorized representative of the group operates the bank account. The deposit kept in the bank is used for giving loans to members for purposes including consumption at the rate of interest decided by the group (usually higher than what the banks charge).

g) Sources of funds are the contribution of members savings, entrance fee, interest from loans, proceeds of joint business operation and income from investment.

Need and Importance of Self-Help Group

- To mobilize the resources of the individual members for their collective economic development.
- To uplift the living conditions of the poor.
- To create a habit of savings, utilization of local resources.
- To mobilize individual skills for group's interest.
- To create awareness about right.
- To assist the members financial at the time of need.
- Entrepreneurship development.
- To identify problems, analyzing and finding solutions in the groups.
- To act as a media for socio-economic development of village.
- To develop linkage with institution of NGOs.
- To organize training for skill development.
- To help in recovery of loans.
- To gain mutual understanding, develop trust and self-confidence.
- To build up teamwork.
- To develop leadership qualities.
- To use it as an effective delivery channel for rural credit.

Characteristics of SHGs

1. They usually create a common fund by contributing their small savings on a regular basis.
2. The groups evolve a flexible system of operations often with the help of the non-governmental organizations (NGOs) and manage their common pooled resource in a democratic manner.

3. Groups consider loan requests in periodical meetings, with competing claims on limited resources being settled by consensus regarding greater needs.
4. Loaning is mainly on the basis of mutual need and trust with minimum documentation and without any tangible security.
5. The amounts loaned are small, frequent and for short duration.
6. Rates of interest vary from group to group depending upon the purpose of loans and are often higher than those of banks but lower than those of moneylenders.
7. At periodical meetings, besides collecting money, emerging rural, social and economic issues are discussed.
8. Defaulters are rare due to group pressure and intimate knowledge of the end use of the credit as also the borrower's economic resources:

Functions of Self-Help Groups

- Enabling members to become self-reliant and self-dependent.
- Providing a forum for members for discussing their social and economic problems.
- Enhancing the social status of members by virtue of their being members of the group.
- Providing a platform for members for exchange of idea.
- Developing and encouraging the decision making capacity of members.
- Fostering a spirit of mutual help and co-operation among members.
- Instilling in members a sense of strength and confidence which they need for solving their problems.
- Providing organizational strength to members.
- Providing literacy and increasing general awareness among members, and
- Promoting numerically and equipping the poor with basic skills required for understanding monetary transactions.

SHGs function on the principle of the five 'P's:

a) Propagator of voluntarism

b) Practioner of mutual help

c) Provider of timely emergency loan

d) Promoter of thrift and savings, and

e) Purveyor of credit.

Norms of SHGs

Any self help group will have certain norms : help to function properly for sustainability of SHGs. It is mandatory for any SHG to have certain byelaws pertaining to:

Objectives of the group:

- **Meetings** - time, periodicity
- **Savings** - amount, periodicity, rate of interest.
- **Credit** - procedure for sanction, ceiling amount, purposes, rate of interest to be charged, repayment period.
- **Fines** - defaulters in attending meetings, savings and credit repayment
- **Leadership** - election or nomination of leaders, rotation of leaders, etc.
- **Personal / Social improvement** - minimum literacy to be achieved, social work to be done, convergence of facilities, etc.

Rules and Regulations of SHGs

I. Memberships

- Only one responsible person from a family aged around 21-60 years can become member in the groups.
- The members/persons who are involved in any party politics activities or involved in any type of antisocial activities or the willful defaulters are not eligible for membership in the group.
- The size of the group shall be around 10 to 20 members.

II. Group Meeting

- The group members meetings should be held once a week or fortnight / month regularly on a convenient day, place and time as decided by the group.

- The members should attend all the meeting in time without fail.
- The member who is unable to attend the meeting for genuine reasons, the same has to be intimated to the group in person or through a messenger in advance or atleast informed at the same meeting. If fails, the member is liable to pay fine for the absence as decided by the group.
- The late comers for the meeting and those who walk out in the middle without intimating the chairpersons are liable to pay fine as decided by the group.
- If a member is absent for three consecutive meetings without genuine reasons such person's membership shall be suspended or cancelled with or without notice.
- Unrelated issues/points should not be brought for discussion in the group meetings.
- The members should not use vulgar words or physical force against any member during the meeting and the violators or mis-behaviours have to pay fine as decided by the group or their membership shall be cancelled.
- The members should not sleep or stay separate from the group during the meeting.
- All the members should have to sign in the minute's book at the end of the meeting after recorded proceedings of each meeting are read out and confirmed.

III. Members Participation in the Group

- All the members should participate in the discussions and decision-making process orally and mentally in the group meetings.
- Equal opportunities and encouragement should be given to all the members for their full participation in the meeting and in all the activities of the group.
- All the members should attend the related training/workshop/seminars/exposures, etc. within and outside the village without fail. The violators are liable to pay fine except for the genuine reasons.

- The members should co-operate and participate in all the developmental activities related to the members families, village, the group environment, etc. Appropriate actions against non-cooperators/non-participators shall be taken by the group.
- The members should participate in researching/learning dissemination and adoption of appropriate technologies for development.
- The illiterate members should show interest and put efforts to become literates.
- The group shall run a learning center at the convenient time for this purpose.
- All the members should involve in regular savings and credit management activities with a special focus.

IV. Duties and Responsibility of the Members

- All the members should promote co-operation and unity in the group.
- The members should create equal opportunities and the give encouragement to all the members in the groups.
- The member should mobilize, use and manage the needy resources properly/judicially. It is the responsibility of all the members to take necessary collective action against the willful defaulters and recovers the loan amount.
- The members should take responsibility carefully for their all-round development and should also take leadership responsibility in the group with service motive.
- All the members should involve in planning, implementing, monitoring and evaluate the development programmes of the family, village and environment time to time and to give attention for the results of evaluations with proper actions.
- The member should promote and protect the unity and integrity of the group and the village.

V. Executive Committee

- An executive committee consisting three representatives selected unanimously in the group should take overall responsibility of the smooth functioning of the group.

- The period of the executive committee shall be six months or one year as group decides and the new committee should be selected three months in advance and trained to take over the position of their representatives.
- The above three position of the representatives shall be called as (a) President, (b) Secretary (c) Treasurer. Two out of the above three shall jointly operate the Group's bank account. This committee is responsible for the group's cash at hand, cash at bank or post office.
- The executive committee members should ensure remitting the cash at hand immediately to the group's bank account. The cash in hand should not be kept more than two days for emergencies, if kept, the concern member should pay fine plus interest as decided by the group.
- The active member/cashier/treasurer should ensure proper maintenance of books of accounts of the group on day-to-day basis regularly up-to-date.
- Responsible for getting the accounts inspected once a month and audited once in six months or year and to submit the accounts statements and reports to the group meeting for appraisal.
- The executive committee should facilitate the programme planning, timely implementation, monitoring, evaluation and actions.
- The executive committee should facilitate the regular group meetings and smooth functioning of the group.
- The executive committee members should have contact with the Govt.
- Departments, credit institutions, voluntary organizations and other related institutions and mobilize resources for the improvement of group and village.

VI. Members Savings in the Group

- Each and every member should save at least minimum or maximum as per members wish per week or per month in the group which should be maintained in each individual members name.
- Savings account may be withdrawn by the members only at critical circumstances with prior approval of the group. However,

the member should maintain a minimum balance of Rs. 1000/- in their savings account.

- No interest will be paid for the members savings with group. But 12% interest shall be paid for the amount kept in the group as fixed deposit for a minimum period of six months.
- The members should save from the family income earned only, but not the borrowed money for interest.
- The members will not be encouraged to adjust their savings amount against their loans due to the group. Only at extreme circumstances, the group shall consider to adjust.

VII. The Credit Management

- The credit can be given to needy members of group for the purposes, such as consumption, income generation, asset creation, clearing the old burden loans, socio-religious and any other appropriate purposes.
- The credit shall be given to needy person of group only after careful study, through discussion and unanimous decision on quantum of credit, rate of interest and repayment schedule for each loan.
- All SHG members regularly save a small amount. The amount may be small have to be regular and continuous habit with the members.
- Savings first credit later should be the motto of every SHG member.
- The loaner should give a written agreement to the group for the loan amount as the official procedures in the presence of the witness to the group at the time of taking credit. The loaner should give a guarantor within the group if necessary.
- The loan amount should be utilized for the agreed purpose only. In case of any change of purpose, it should be approved by the group in advance. The violators are liable to pay fine/penal rate of service charge.
- The loaner should repay the credit/loan amount with service charge as per repayment schedule agreed upon. If failed, such member is liable to pay fine or penal rate of service charge

@ Rs. 12% annum in addition to the normal rate of service charge from the date of overdue or as decided by the group.

- The members those who are irregular for the SHG meetings or irregular in savings are not eligible for credit/loan from the group.
- All the members shall have equal opportunity for loan from the group on eligibility/priority basis.
- The funds of the group should be revolved to the optimum extent for the benefit of all the members.

SHG Linkage Models

1. **Direct linkage model**

In case of the direct linkage model the bank identifies the group (or facilitates evaluation of the group) and deals with the SHG directly for both mobilizing savings and for making available credit to as a whole or to individual members. Group members act as collateral security. In this model the credit is generally made available to the group and members to be financed are identified by the group itself which takes the responsibility of loan repayment.

2. **Modified Linkage Model**

Here the activity and member to whom loan is given is identified by group. The group is morally responsible for repayment but credit is given as any individual loan.

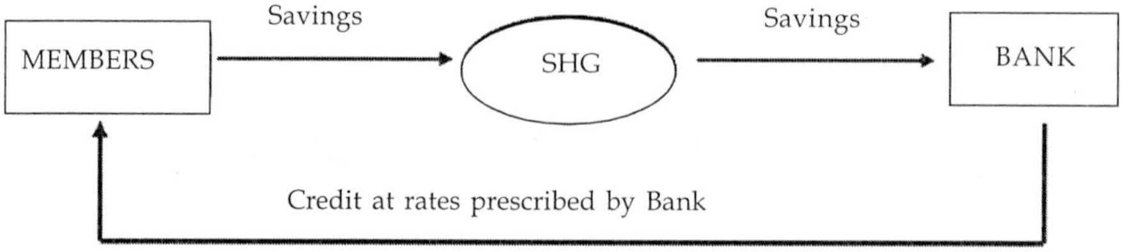

3. NGO Model

In this model NGO is not the financial intermediary. The NGO's role is only in group evolution and stabilization, where as the financial linkage is directly with the group. In this model NGO is not the financial intermediary. The NGO's role is only in group evolution and stabilization, whereas the financial linkage is directly with the group.

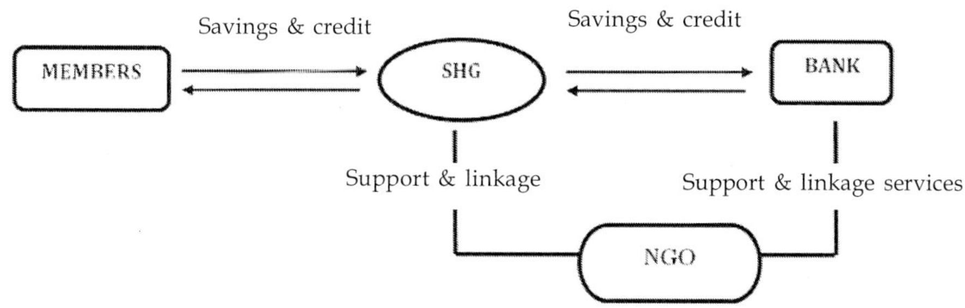

4. Modified NGO Model

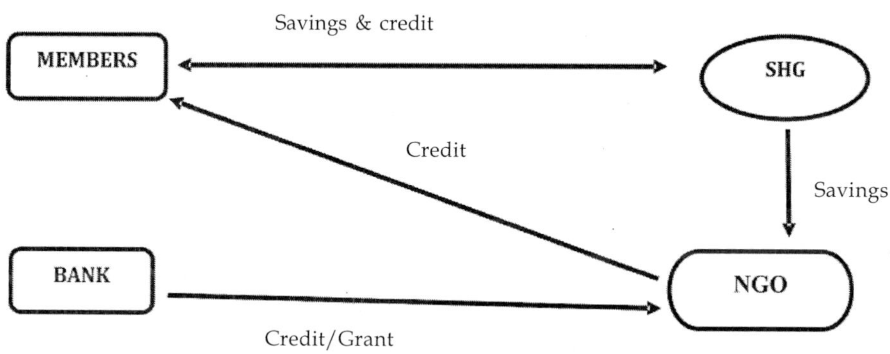

5. Indirect NGO Model

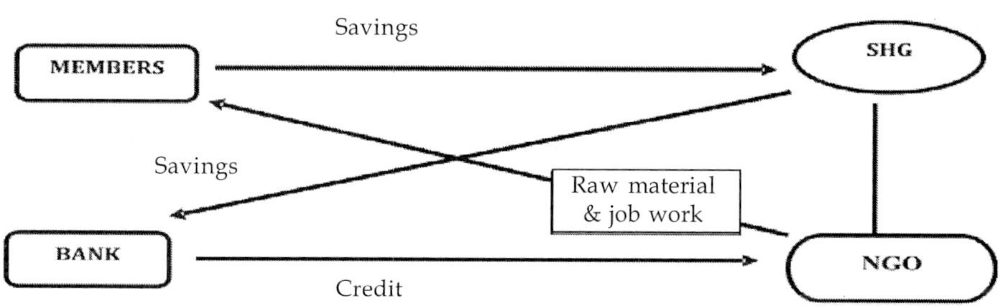

6. **IFAD (International Fund for Agricultural Development) Model**

In this model, the NGO is involved as in (iii) above but the line departments of Government like Women Development Corporations, Sericulture, Rural Development are involved in identification of activity, beneficiary, etc. The model is in existence in areas where IFAD projects are being implemented like Tamil Nadu, Maharashtra and Uttar Pradesh.

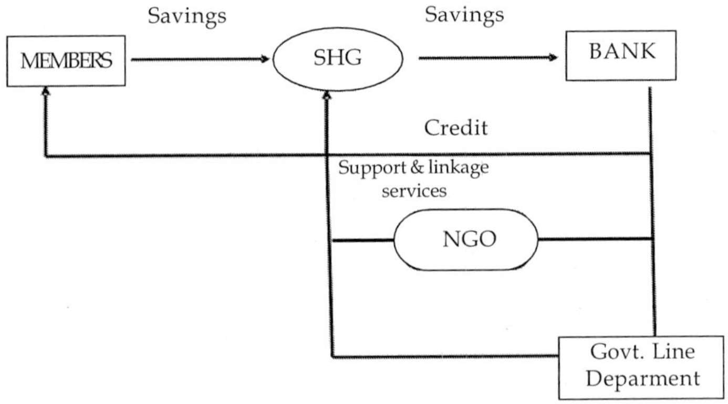

7. **SGSY Model**

The most important feature of this model is the addition of Monitoring authority that is District Rural Development Authority (DRDA). It monitors for all the five like Members, Group, Bank, Department and NGO. Processing of the model is like this, members saves and submit their savings in the name of a Group then Group deposit the amount of saving to banks where banks, in turn, lends to the group directly. Here NGO and concerned department associated with SHG, Bank and DRDA and Members, DRDA and Banks respectively with their promotional and support services.

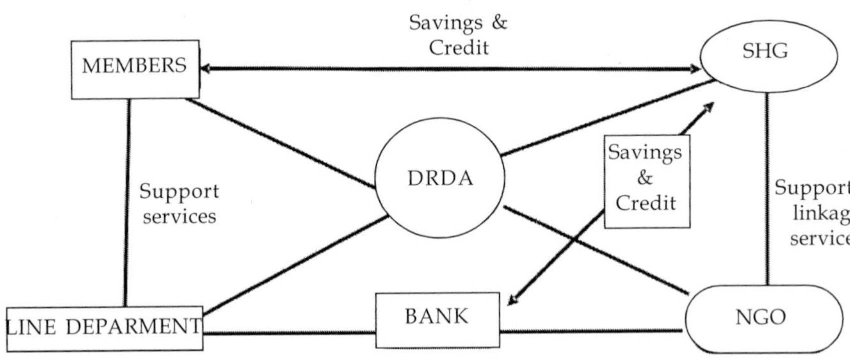

Stages in SHG formation

- as given by **Tuckman**

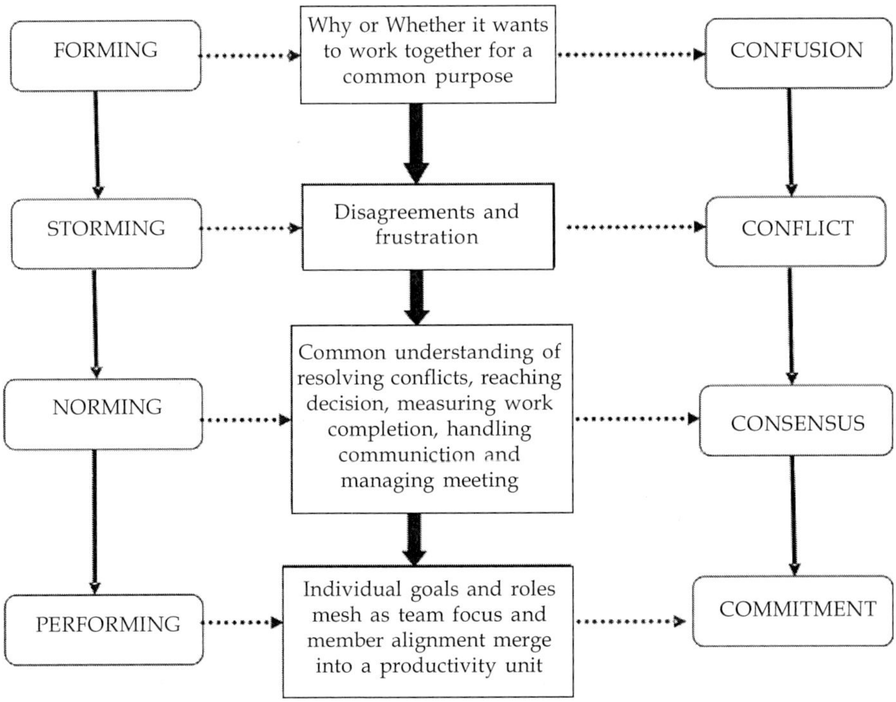

According to NABARD, five stages of SHG development are:

Sl.	Stages of development	Role of NGO/ SHPI	Time Period	Focus of Activities
1	Pre-formation	Initiator	1-2 months	Identifying poor through PRA
2	Formation	Facilitator	3-6 months	Motivation to form groups, develop rules and norms, conducting meetings, pooling savings, lending and collection of small loans, adjustment
3	Stabilization (Phase-I)	Advisory /Managerial	7-12 months	Leadership stabilization, training of leaders, increased savings and handling of group level transactions

4	Stabilization (Phase-II)	Advisory /Managerial	13-18 months	Initiation of income generating activities , linkage of banks, support to new groups
5	Growth and expansion	Advisory /Managerial or Institution building	19 months	Strength of linkages with banks, creation of assets for groups and promotion of new groups

Note : NGO- Non-governmental organisations

SHPI- Self-Help Promoting Institutions

SHG - Conceptual framework and key features

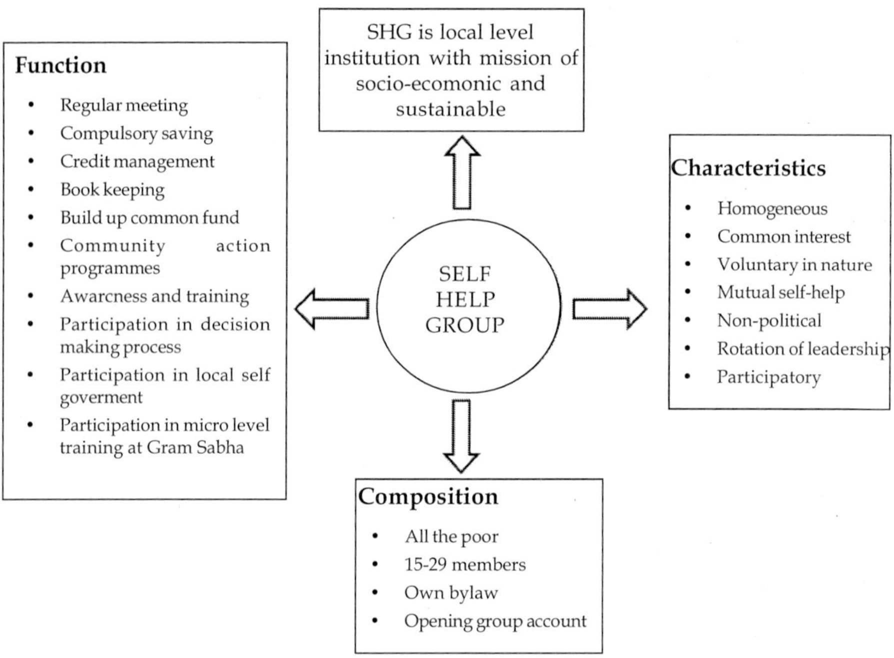

e-Shakti – A Project of NABARD on digitization of Self-Help Groups

e-Shakti or Digitisation of SHGs is an initiative of Micro Credit and Innovations Department of NABARD in line with our Hon'ble PM Narendra Modi statement, 'we move with the dream of electronic digital India...'. Digital India is a Rs. 1.13-lakh crore initiative of

Government of India to integrate the government departments and the people of India and to ensure effective governance. It is to "transform India into digital empowered society and knowledge economy".

Keeping in view the Government of India's mission for creating a digital India, NABARD launched a project for digitisation of all Self-Help Group (SHG) in the country. The project is being implemented in 100 districts across the country.

Definition: e-Shakti consists of two word E-Electronic and Shakti-Power. It means empower. A project on Digitisation of Self-Help Groups' (SHGs) financial and non-financial records initiated by NABARD for various stakeholders who are working in the field of microfinance.

Aims and Objectives

The project aims at digitisation of all the SHG accounts to bring SHG members under the fold of Financial Inclusion thereby helping them access wider range of financial services together with increasing the bankers' comfort in credit appraisal and linkage by way of:

- Integrating SHG members with the National Financial Inclusion agenda;
- Improving the quality of interface between SHG members and Banks for efficient and hassle free delivery of banking services by using the available technology;
- Facilitate convergence of delivery system with SHGs using Aadhaar linked identity.

The Background

The need of digitisation of records of SHGs has been felt for quite some time due to patchy and delay in maintenance of books of accounts. Transparent and proper maintenance of records of SHGs will facilitate in nurturing and strengthening of SHGs. It is also felt that digital empowerment will help in bringing SHGs on a common web based e-platform by making book keeping easy for low literacy clients. This will help in promoting national agenda of Financial Inclusion and pave the way of credibility of SHG data which can later be used by Credit Bureaus to reduce the issues related to multiple financing by banks.

Stakeholders of the Project

Rural poor community and members of SHGs, Self-Help Promoting Institutions (SHPI), Non-Government Organisations (NGOs), Banks, NABARD, Government development departments and agencies like SLRM, NLRM, etc. At a later stage Credit Bureaus may also be on board.

Attributes of the project

- e-book keeping for the SHGs
- Regular updates of transactional data
- Reports generated in the formats as required by stakeholders like bankers
- Inbuilt automatic grading of SHGs based on NABARD/IBA(for NRLM) norms
- Auto generation of Loan application for the bankers on input of resolution to borrow by SHG.

The project will also help in:

- A comprehensive information base and robust MIS can be developed about poor community covered, which may facilitate suitable interventions and convergence of other programme for social and financial empowerment;
- It will help in identifying suitable interventions and support for proper nurturing and strengthening of SHGs.
- Ease of transfer of social benefits and Direct Benefit Transfer (DBT) through Aadhaar linked accounts and convergence with other Government benefits.

The USP of e-shakti software is 'one-click' availability of social and financial information of all the members of the Self-Help Groups.

Components of the Project

- Mapping of existing SHGs in the district (bank wise, branch wise);
- Training of volunteers to collect SHG wise/member wise data;

- Data feeding through a customised software in central server;
- Hosting of data on the web under a dedicated website i.e. www.eshakti.nabard.org ;
- Maintaining data centre and data recovery centre;
- Regular update of the transactional data
- Generating MIS for various users.

Process

Information of all the SHGs and their members are uploaded onto the website. The transactions are updated through 'apps' on android Mobile/Tablet. Data authenticity is ensured through SMS alerts to members and sample audits. MIS reports on groups are generated and progress is tracked on a real time basis. Overall, the whole eco-system was designed to address the complex issue related to patchy financial records of SHGs.

Information captured

@Member Level

Name, address, gender, marital status, Physically Challenge status, Aadhaar details, Voter ID card details, mobile number, BPL/ APL status, membership of any Joint Liability Group (JLG), house type, availability of toilets, electricity connection and other financial details like saving bank account number, savings, borrowing & repayment, life/medical insurance, micro pension policy (if any).

@SHG level

Name, address, date of formation, name of SHPI/NGO, programme under which supported, savings habits, lending policy, bank linkage, details of periodic savings collected and internal lending, utilisation of bank credit availed for members.

Output and MIS

- Member wise details of SHGs on saving, lending, attendance;
- SHG and member wise credit history;
- Financial statement of SHGs - Balance sheet and Profit and Loss account;

- Grading chart of SHGs;
- Micro Credit Plan of the SHGs;
- Audit report;
- Bank linkage details - savings and credit disbursement;
- Other periodical MIS on performance of SHGs.

Impact

- Provided credit to SHGs based on real time performance;
- Significantly reduced Saving-Credit linkage gap;
- Captured credit history of members;
- Promoted transparency through real time SMS alerts in 10 languages to members;
- Aided convergence of SHGs with other Government Programmes;
- Integrated members with the larger Financial Inclusion Agenda.

Paving way for Credit Linkage

- Bank branches to access the portal through Intranet or web link using their branch IFSC and password;
- More than 31 MIS reports like Savings, meeting, Credit linkage, Repayment and Demand Collection Balance can be generated sitting in the Bank branch itself;
- Easier processing of loan using EShakti system generated prefilled application forms with all the members and group information.

Challenges

- Sourcing of information from poor database and records;
- Large scale training and capacity building of SHGs, SHPIs and others involved in implementation of the programme;
- Capture of field level information from SHGs in a limited time and periodic upload of savings and credit details of SHGs;
- Co-operation from banks;

- GPRS connectivity;
- Initial funding of the pilot is being met by NABARD. There will be a huge fund requirement for scaling this pilot from the present level to 86 lakh SHGs across the country.

❒❒❒

Chapter 14

Women Develpment Programmes and National Policies

Child Protection & Welfare Schemes

1. Rajiv Gandhi Scheme for Empowerment of Adolescent Girls (RGSEAG) Sabla
2. Kishori Shakti Yojana
3. Integrated Child Protection Scheme (ICPS)
4. Rashtriya Bal Kosh (National Children's Fund)

Rajiv Gandhi Scheme for Empowerment of Adolescent Girls (RGSEAG) Sabla

The Rajiv Gandhi Scheme for Empowerment of Adolescent Girls (RGSEAG) SABLA is a centrally sponsored program of **Government of India** initiated on April 1, 2011 under **Ministry of Women and Child Development**.

Objective

The objectives of the program are:

- Enable the Adolescent girls for self-development and empowerment
- Improve their nutrition and health status.
- Promote awareness about health, hygiene, nutrition, adolescent reproductive and sexual health (ARSH) and family and child care.
- Upgrade home-based skills, life skills and integrate with the National Skill Development Program (NSDP) for vocational skills.

- Mainstream out of school adolescent girls into formal/non formal education.
- Provide information/guidance about existing public services, such as PHC, CHC, Post Office, Bank, Police Station, etc.

Content

An integrated package of services is to be provided to adolescent girls as follows:

- Nutrition provision
- Iron and Folic Acid (IFA) supplementation
- Health check-up and Referral services
- Nutrition & Health Education (NHE)
- Counseling/Guidance on family welfare, ARSH, child care practices and home management
- Life Skill Education and accessing public services
- Vocational training for girls aged 16 and above under NSDP

Eligibility criteria

The program would cover adolescent girls (only covered out of school girls)11–18 years old under all ICDS projects in selected 200 districts in all states/UTs in the country. The target group would be subdivided into 11-15 and 15–18 years.

Kishori Shakti Yojana

Kishori Shakti Yojana (KSY) seeks to empower adolescent girls, so as to enable them to take charge of their lives. This scheme is a redesign of the already existing Adolescent Girls (AG) Scheme being implemented as a component under the centrally sponsored Integrated Child Development Services (ICDS) Scheme.

The objectives of the Scheme are as follows:

i) To improve the nutritional and health status of girls in the age group of 11-18 years;

ii) To provide the required literacy and numeracy skills through the non-formal stream of education, to stimulate a desire for more social exposure and knowledge and to help them improve their decision making capabilities;

iii) To train and equip the adolescent girls to improve/ upgrade home-based and vocational skills;

iv) To promote awareness of health, hygiene, nutrition and family welfare, home management and child care, and to take all measure as to facilitate their marrying only after attaining the age of 18 years and if possible, even later;

v) To gain a better understanding of their environment related social issues and the impact on their lives; and

vi) To encourage adolescent girls to initiate various activities to be productive and useful members of the society.

Target group

- All eligible girls in the age group of 11-18 years, irrespective of income levels of the family, are eligible to receive assistance under the scheme.
- Preference is given to younger girls, in the age group of 11-15 years, and belonging to poor families.
- 20 girls in the age group of 11-18 years are identified from all the eligible girls in this age group in the Anganwadi Centres.
- These 20 girls are enrolled for a period of six months in the balika mandals. In each ICDS project, only 10%of the total Anganwadi Centres are selected on rotation to serve as balika mandals.

Integrated Child Protection Scheme (ICPS)

The Integrated Child Protection Scheme is expected to significantly contribute to the realization of Government/State responsibility for creating a system that will efficiently and effectively protect children. It is based on cardinal principles of - protection of child rights and best interest of the child. Hence, the ICPS objectives are: to contribute to the improvements in the well being of children in difficult circumstances, as well as to the reduction of vulnerabilities to situations and actions that lead to abuse, neglect, exploitation, abandonment and separation of children.

These will be achieved by: (i) improved access to and quality of child protection services; (ii) raised public awareness about the reality of child rights, situation and protection in India; (iii) clearly articulated responsibilities and enforced accountability for child protection (iv) established and functioning structures at all government levels for delivery of statutory and support services to children in difficult circumstances; (v) introduced and operational evidence based monitoring and evaluation.

Target Groups: The ICPS will focus its activities on children in need of care and protection and children in conflict and contact with the law.

Specific Objectives

A. To institutionalize essential services and strengthen structures:

i) Establish and strengthen a continuum of services for emergency outreach, institutional care, family and community based care, counselling and support services;

ii) Put in place and strengthen necessary structures and mechanisms for effective implementation of the scheme at the national, regional, state and district levels;

iii) Define and set standards of all services including operational manuals for the functioning of statutory bodies.

B. To create database and knowledge base for child protection services:

i) Create mechanisms for a child protection data management system including MIS and child tracking system in the country for effective implementation and monitoring of child protection services;

ii) Undertake research and documentation.

C. To strengthen child protection at family and community level:

i) Build capacities of families and community to strengthen care, protection and response to children;

ii) Create and promote preventive measures to protect children from situations of vulnerability, risk and abuse.

D. To ensure appropriate inter-sectoral response at all levels:

Coordinate and network with all allied systems i.e. Government departments and Non-Government agencies providing services for children for effective implementation of the scheme.

E. To raise public awareness:

i) Educate public on child rights and protection

ii) Raise public awareness at all levels on situation and vulnerabilities of children and families

iii) Inform the public on available child protection services, schemes and structures at all levels

Rashtriya Bal Kosh (National Children's Fund)

The National Children's Fund (NCF) was set up by the Government of India during the year 1979. The National Policy for Children, announced in the year 1974 recognises children of the Nation as a 'Supremely important asset'. Their nurture and development are responsibility of the nation. In order to help and promote various welfare and development programmes for children, the Government of India, in the International Year of the Child (1979), created National Children's Fund (NCF) with a corpus fund of rupees one lakh under the Charitable Endowment Act, 1890. The Minister of State, Ministry of Women and Child Development is the Chairperson of Board of Management of NCF. The Secretary, Ministry of Women and Child Development is its Working Chairperson.

The objectives of NCF are to raise funds from individuals, institutions, Corporates and others, to promote and fund the various programmes for children who are affected by natural calamities, disasters, distress and in difficult circumstances through voluntary agencies and State Governments in unserved and underserved areas including tribal and remote areas (as per the National Charter for Children, 2003) and children of prisoners, children affected by riots, aggression, trafficking and children of prostitutes, and to implement various programmes. The secretariat of the fund is located in National Institute of Public Co-operation and Child Development (NIPCCD), an autonomous body under the aegis of the Ministry of Women and Child Development.

Women Empowerment Schemes

1. Beti Bachao, Beti Padhao Scheme
2. One Stop Centre Scheme
3. Women Helpline Scheme
4. UJJAWALA : A Comprehensive Scheme for Prevention of trafficking and Rescue, Rehabilitation and Re-integration of Victims of Trafficking and Commercial Sexual Exploitation
5. Working Women Hostel
6. Rajiv Gandhi National Creche Scheme For the Children of Working Mothers
7. SWADHAR Greh (A Scheme for Women in Difficult Circumstances)
8. Support to Training and Employment Programme for Women (STEP)
9. Nari Shakti Puraskar
10. Archived Maternity Benefit Programme
11. Mahila police Volunteers
12. Mahila E-Haat

Beti Bachao, Beti Padhao Scheme

Beti Bachao, Beti Padhao (BBBP) scheme was started on January 22, 2015 as a joint initiative of the Ministry of Women and Child Development, Ministry of Health and Family Welfare and Ministry of Human Resource Development under coordinated and convergent efforts to empower the girl child. First, it was initiated in 100 districts with low girl child sex ratio. Based on the child sex ratio, 100 districts were selected as pilot districts along with at least one district in each of the states/union territories covered by the 2011 census. In Haryana, Olympic bronze medallist Sakshi Malik was made the brand ambassador for 'Beti Bachao, Beti Padhao scheme in August 2016.

The Aims and Objectives

The complete aim of Beti Bachao, Beti Padhao scheme is to celebrate the birth of a girl child to prevent the violation of the interests of girls as a result of outdated and conservative thoughts. This plan was launched for the following objectives for girls' education and welfare:

i) **To stop the discrimination of the girl child and practice of sex determination test**

Today, the female sex ratio is decreasing on an alarming level in Asia. Our country is at the top of this declining ratio. Under the Beti Bachao, Beti Padhao scheme, mainly female and male sex ratio has been focused upon and major steps are being taken towards the prevention of gender discrimination.

ii) **To ensure the survival and protection of girls**

In our country everyday, you can read the news in the newspapers that a female embryo, an unborn baby girl, was found dead in the dust-bin, wrapped in newspapers or near the railway station, etc. What is happening in our country? This shows somewhere some sick mindset prevails in our society. Beti Bachao, Beti Padhao scheme is a huge step in the direction of stopping this practice and ensuring the existence and safety of every child.

iii) **To ensure the participation of girls in education and other areas**

Save the girl child and ensure her safety with a view to strengthen and create a better India. According to Prime Minister Narendra Modi, every girl child should be educated in this country so that she can learn what she wants.

The key features

There are two major features of Beti Bachao, Beti Padhao (BBBP) scheme:

i) **Mass campaign:**

The aim of the campaign is to ensure that the birth and upbringing of the girl child lead to their empowerment so that she becomes a proud citizen of the country without discrimination. The campaign has been launched at the community level in 100 districts along with national, state and district levels with immediate effect.

ii) **Multisectoral action to cover all states and union territories in 100 selected districts with CSR (as a pilot scheme)**

Ministry of Human Resource Development and Ministry of Health and Family Affairs have taken joint steps to ensure the survival and protection of girl's education. These involve multisectoral, united efforts for the management of BBBP by all departments at the level of district Collectors/Deputy Commissioners (DC).

Steps by the Ministry of Women and Child Development

- To promote the registration of pregnancies in the first trimester in Anganwadi centres;
- Undertake training of new women entrepreneurs;
- Community mobilization and sensitization;
- The participation of gender support;
- Award and recognition of front-running workers and institutions.

Steps by the Ministry of Health and Family Welfare

- Monitor the implementation of Preconception and Prenatal Diagnostic Techniques Act (PCPNDT) Act, 1994
- Increase in institutional deliveries
- Registration of births
- Strengthening PNDT Cells
- Establishment of monitoring committees

Steps by the Ministry of Human Resource Development

- Universal enrolment of girls
- Reducing the dropout rate among girls
- Friendly, amiable behaviour with girls in schools
- To enforce Right to Education (RTE) Act
- Creating functional toilets for girls

How to apply for Beti Bachao, Beti Padhao Scheme

Opening a bank account in the name of girl is the first basic step. It is necessary to comply with the rules to draw the scheme's benefits:

Age limit for Beti Bachao, Beti Padhao Scheme

- All the girls upto 10 years of age are eligible under this scheme. It is necessary to open bank account in their name.

Tax Free scheme - This welfare scheme launched by the Prime Minister has been set kept completely tax free. No amount will be deducted from your account after opening it.

Necessary documents for Beti Bachao, Beti Padhao Scheme

- make sure you have the following documents to open an account:

- Child's birth certificate
- Parent or legal guardian's identity proof
- Proof of address of parent or legal guardian
- Note: - The scheme does not apply to NRIs

As per RBI guidelines, rules and regulations, NRIs cannot apply for this scheme. Sukanya Devi scheme is regulated by Post Office Savings Account Rules, 1981. NRIs, therefore, do not come under this category.

How much benefits girls would derive

Rs. 150 crore will be spent by the Government of India to increase the safety of women in cities. BBBP has been functioning under Ministries of Women and Child Development, Health and Family Welfare and Human Resource Development. Prime Minister Modi has described it as a big step and a boon for the society. The Union budget has allocated Rs. 50 crore for the schemes to protect the safety of women under the Ministry of Road Transport and Highways.

The benefits for the girl child are:

- You would be able to open an account for your girl child which will reduce your financial burden and the girl will get the money for her small needs.
- The government provides the highest interest rate under the BBBP scheme for all small savers. With it, you can save more money for your daughter in the future.

- This account is exempted under the Act, 1961 u / s 80C. The girl's account will be tax-free. This means that no money shall be deducted from the account in the form of tax.

The best part of this plan - The account will mature when the girl reaches 21 years from the time of opening the account. Only at the age of 18, she will receive money for higher education. When she is 21, you will be able to take money from the account for her marriage. The account has a maximum duration of 21 years.

The aim of BBBP scheme is to explain to the people that girls' education and marriage is not a burden to parents. You can arrange for the marriage of your daughter through the money saved under this account. This scheme provides complete financial security to the girl child. After the daughter completes 21 years from the time of opening the account the entire amount will be credited to her account after adding all interest into it.

Strategies

Strategies employed to successfully carry out the scheme are:

- Implement a sustained Social Mobilization and Communication Campaign to create equal value for the girl child and promote her education.
- Place the issue of decline in CSR/SRB in public discourse, improvement of which would be an indicator for good governance.
- Focus on Gender Critical Districts and Cities.

One Stop Centre Scheme

This Scheme of One Stop Centre for Women is implemented since 1st April, 2015 with total project cost of Rs. 18.58 is funded under Nirbhaya fund.

One Stop Centres (OSC) are intended to support women affected by violence, in private and public spaces, within the family, community and at the workplace. Women facing physical, sexual, emotional, psychological and economic abuse, irrespective of age, class, caste, education status, marital status, race and culture will be facilitated with support and redressal. Aggrieved women facing any kind of violence due to attempted sexual harassment, sexual assault, domestic violence, trafficking, honour related crimes, acid attacks or witch-hunting who have reached out or been referred to the OSC will be provided with specialized services.

Funding Pattern: This is a centrally sponsored Scheme funded through Nirbhaya Fund. The Central Government will provide 100% financial assistance to the State Government /UT Administrations under the Scheme. The day-to-day implementation and administrative matters would be the responsibility of the State Government.

The objectives of the Scheme are:

i) To provide integrated support and assistance to women affected by violence, both in private and public spaces under one roof.

ii) To facilitate immediate, emergency and non-emergency access to a range of services including medical, legal, psychological and counselling support under one roof to fight against any forms of violence against women.

Target Group : The OSC will support all women including girls below 18 years of age affected by violence, irrespective of caste, class, religion, region, sexual orientation or marital status. For girls below 18 years of age, institutions and authorities established under Juvenile Justice (Care and Protection of Children) Act, 2000 and the Protection of Children from Sexual Offences Act, 2012 will be linked with the OSC.

Services offered in OSCs

The Centres will be integrated with a Women Helpline to facilitate access to following services:

- **Emergency Response and Rescue Services** - OSC will provide rescue and referral services to the women affected by violence. For this, linkages will be developed with existing mechanisms, such as National Health Mission (NHM), 108 service, police (PCR Van) so that the woman affected by violence can either be rescued from the location and referred to the nearest medical facility (Public/ Private) or shelter home.
- **Medical assistance** - Women affected by violence would be referred to the nearest Hospital for medical aid/examination which would be undertaken as per the guidelines and protocols developed by the Ministry of Health and Family Welfare.

- **Assistance to women in lodging FIR /NCR/DIR**
- **Psycho-social support/ counselling** - A skilled counsellor providing psycho - social counselling services would be available on call. This counselling process will give women confidence and support to address violence or to seek justice for the violence perpetuated. Counsellors shall follow a prescribed code of ethics, guidelines and protocols in providing counselling services.
- **Legal aid and counselling** - To facilitate access to justice for women affected by violence, legal aid and counselling would be provided at OSC through empanelled Lawyers or National/ State/District Legal Service Authority. The aggrieved woman would be provided with an advocate of her choice in case she wants to engage the same to assist the State Prosecutors in trying her case. It would be the responsibility of the Lawyer /Prosecutor to simplify legal procedures for the aggrieved woman and advocate for her exemption from court hearings. In case the trial or inquiry relates to an offence of rape as defined under section 376, 376A - D IPC, it would be the duty of the Prosecutors trying the case to complete the inquiry or trial as far as possible within a period of two months from the date of filing of charge sheet.
- **Shelter** - The OSC will provide temporary shelter facility to aggrieved women. For long term shelter requirements, arrangements will be made with Swadhar Greh/Short Stay Homes (managed/affiliated with government/NGO). Women affected by violence along with their children (girls of all ages and boys up till 8 years of age) can avail temporary shelter at the OSC for a maximum period of 5 days. The admissibility of any woman to the temporary shelter would be at the discretion of Centre Administrator.
- **Video Conferencing Facility** - To facilitate speedy and hassle free police and court proceedings the OSC will provide video conferencing facility (through Skype, Google Conferencing, etc.). Through this facility if the aggrieved woman wants, she can record her statement for police/ courts from OSC itself using audio - video electronic means as prescribed under sections 161(3), 164(1) and 275(1) of the Code of Criminal Procedure and section 231(1) in line with Order XVIII Rule 4 of the Code of Civil Procedure. This facility will be provided only after consultation among Superintendent of Police, District and Sessions Judge of the concerned district (place of incident).

Women Helpline Scheme

The Scheme of Universalisation of Women Helpline is intended to provide 24 hours immediate and emergency response to women affected by violence through referral (linking with appropriate authority such as police, One Stop Centre, hospital) and information about women related government schemes programs across the country through a single uniform number. This scheme is being implemented since from 01-04-2015 by the Union Ministry of Women and Child Development.

The objectives of the Scheme are

- To provide toll-free 24-hours telecom service to women affected by violence seeking support and information.
- To facilitate crisis and non-crisis intervention through referral to the appropriate agencies such as police/Hospitals/Ambulance services/District Legal Service Authority (DLSA)/Protection Officer (PO)/OSC.
- To provide information about the appropriate support services, government schemes and programmes available to the woman affected by violence, in her particular situation within the local area in which she resides or is employed.

Target Group: Any woman or girl facing violence within public or private sphere of life or seeking information about women related programmes or schemes.

How to Apply

The Women Helpline will be up and active 24 hours a day 7 days a week to any woman or girl suffering violence or in distress in the following manner:

1. Telephone - landlines, mobile phones through calls, SMS/text messaging, mobile apps and fax messages are considered.
2. Internet - Via emails, web-posts, web-interface, social networking sites i.e. web page, facebook, twitter, mygov, etc. can be considered.

Call to any of the following Women Helpline Nos.

1. Call to Women's Helpline: 181 Toll free

2. Call to 1091- Police Helpline, 102 (hospital/Ambulance) & 108 (Emergency Response Services)
3. Contact to One Stop Centre (OSC)

Ujjawala: A Comprehensive Scheme for Prevention of trafficking and Rescue, Rehabilitation and Re-integration of Victims of Trafficking and Commercial Sexual Exploitation

Effective since 1st April, 2016. The new scheme has been conceived primarily for the purpose of preventing trafficking on the one hand and rescue and rehabilitation of victims on the other.

Objectives of the scheme

- To prevent trafficking of women and children for commercial sexual exploitation through social mobilization and involvement of local communities, awareness generation programmes, generate public discourse through workshops/seminars and such events and any other innovative activity.
- To facilitate rescue of victims from the place of their exploitation and place them in safe custody.
- To provide rehabilitation services both immediate and long-term to the victims by providing basic amenities/needs, such as shelter, food, clothing, medical treatment including counselling, legal aid and guidance and vocational training.
- To facilitate reintegration of the victims into the family and society at large.
- To facilitate repatriation of cross-border victims to their country of origin.

Target group/beneficiaries

- Women and children who are vulnerable to trafficking for commercial sexual exploitation.
- Women and children who are victims of trafficking for commercial sexual exploitation.

Implementing agencies: The implementing agencies can be the Social Welfare/Women and Child Welfare Department of State Government, Women's Development Corporations, Women's Development Centres,

Urban Local Bodies, reputed Public/Private Trust or Voluntary Organizations. The organization must have adequate experience in the field of trafficking, social defence, dealing with women and children in need of care and protection, children in conflict with law, etc.

Working Women Hostel

With the progressive change in the socio-economic fabric of the country more and more women are leaving their homes in search of employment in big cities as well as urban and rural industrial clusters. One of the main difficulties faced by such women is lack of safe and conveniently located accommodation. The Government of India being concerned about the difficulties faced by such working women, introduced a scheme in 1972-73 of grant-in- aid for construction of new/ expansion of existing buildings for providing hostel facilities to working women in cities, smaller towns and also in rural areas where employment opportunities for women exist. Based on an evaluation, the existing scheme has been revised to promote availability of safe and conveniently located accommodation for working women who need to live away from their families due to professional commitments.

Objectives: The objective of the scheme is to promote availability of safe and conveniently located accommodation for working women, with day care facility for their children, wherever possible, in urban, semi urban, or even rural areas where employment opportunity for women exist. To achieve this objective, the scheme will assist projects for construction of new hostel buildings, expansion of existing hostel buildings and hostel buildings in rented premises. The working women's hostel projects being assisted under this scheme shall be made available to all working women without any distinction with respect to caste, religion, marital status, etc., subject to norms prescribed under the scheme. While the projects assisted under this scheme are meant for working women, women under training for job may also be accommodated in such hostels subject to the condition that taken together, such trainees should not occupy more than 30% of the total capacity the hostel and they may be accommodated in the hostels only when adequate numbers of working women are not available. Children of working women, up to the age of 18 years for girls and up to the age of 5 years for boys may be accommodated in such hostel with their mothers.

Beneficiaries: Following categories of working women and their children will be covered under this Scheme:

i) Working women, who may be single, widowed, divorced, separated, married but whose husband or immediate family does not reside in the same city/area. Particular preference may be given to women from disadvantaged sections of the society. There should be also provision for reservation of seats for physically challenged beneficiaries.

ii) Women who are under training for job provided the total training period does not exceed one year. This is only on the condition that there is vacancy available after accommodating working women. The number of women under training for job should not exceed 30% of the total capacity.

iii) Girls up to the age of 18 years and boys upto the age of 5 years, accompanying working mothers will be provided accommodation, with their mothers. Working mothers may also avail of the services of the Day Care Centre, as provided under the scheme.

Rajiv Gandhi National Creche Scheme for the Children of Working Mothers

The Rajiv Gandhi National Creche Scheme for the children of working mothers is the scheme launched by central government (Ministry of Woman and Child Development). The main objective of this scheme is to provide a nursery where babies and young children are cared for during the working day. Employment of women has resulted in increased opportunities for their employment, and more and more women are now in gainful employment, working within or outside their homes. Thus the children of these women, who were earlier getting support from relatives and friends while their mothers were at work, are now in need of day care services which provide quality care and protection for the children. Therefore women need a safe place for their children in their absence. It has become necessary to provide support to the young children in terms of quality, substitute care, and other services while the mothers are at work. Effective day care for young children is essential and a cost-effective investment as it provides support to both mothers and young children.

Benefits of Rajiv Gandhi National Creche Scheme:

- Rajiv Gandhi National Creche scheme will provide day-care facilities for children (6 months to 6 years) of working mothers in the community
- Improved nutrition and health status of children
- Physical, mental, emotional and social development of children will be there
- Educate and empower parents /caregivers for better childcare
- Sleeping facility for children
- Health check and immunization facility which will take care of health of child
- Supplementary nutrition will be provided for better growth
- Early stimulation for children below 3 years and pre-school education for 3 to 6 years old children
- Food which contains rich nutrients will be provided to babies.

Eligibility for Rajiv Gandhi National Creche Scheme:

1. The child should be 6 months to 6 years of age
2. Working women in rural and urban areas who are employed for a minimum period of 15 days in a month, or six months in a year

Document required getting enrol into Creche:

1. Childbirth certificate is required
2. Identity card of particular organization where child's mother is working or such any document which will show that the woman is working into particular organization since last 6 months

A user is required to pay following charges:

1. BPL families - Rs 20/- per child per month
2. Families with Income (Both Parents) of upto Rs. 12,000/- per month - Rs 100/- per child per month
3. Families with Income (Both Parents) of above Rs. 12,000/- per month - Rs 200/- per child per month

Swadhar Greh (A Scheme for Women in Difficult Circumstances)

It was launched by the Department of Women and Child Development in 2001- 02. The scheme through the provisions of shelter, food, clothing, counselling, training, clinical and legal aid aims to rehabilitate such women in difficult circumstance.

Vision: The scheme envisions a supportive institutional framework for women victims of difficult circumstances so that they could lead their life with dignity and conviction. It envisages that shelter, food, clothing, and health as well as economic and social security are assured for such women. It also envisions that the special needs of these women are properly taken care of and under no circumstances they should be left unattended or abandoned which could lead to their exploitation and desolation.

Objectives

Under the Scheme, Swadhar Greh will be set up in every district with capacity of 30 women with the following objectives:

1. To cater to the primary need of shelter, food, clothing, medical treatment and care of the women in distress and who are without any social and economic support.
2. To enable them to regain their emotional strength that gets hampered due to their encounter with unfortunate circumstances.
3. To provide them with legal aid and guidance to enable them to take steps for their readjustment in family/society.
4. To rehabilitate them economically and emotionally.
5. To act as a support system that understands and meets various requirements of women in distress.
6. To enable them to start their life afresh with dignity and conviction.

For big cities and other districts having more than 40 lakh population or those districts where there is a need for additional support to the women, more than one Swadhar Greh could be established. The capacity of Swadhar Greh could be expanded up to 50 or 100 on the basis of need assessment and other important parameters.

Beneficiaries

The benefits of the component could be availed by women above 18 years of age of the following categories:

- Women who are deserted and are without any social and economic support;
- Women survivors of natural disasters who have been rendered homeless and are without any social and economic support;
- Women prisoners released from jail and are without family, social and economic support;
- Women victims of domestic violence, family tension or discord, who are made to leave their homes without any means of subsistence and have no special protection from exploitation and/ or facing litigation on account of marital disputes; and
- Trafficked women/girls rescued or runaway from brothels or other places where they face exploitation and Women affected by HIV/AIDS who do not have any social or economic support. However such women/ girls should first seek assistance under UJJAWALA Scheme in areas where it is in operation.

Women affected by domestic violence could stay upto one year. For other categories of women, the maximum period of stay could be up to 3 years. The older women above the 55 years of age may be accommodated for maximum period of 5 years after which they will have to shift to old age homes or similar institutions.

Swadhar Greh facilities could also be availed by the children accompanying women in the above categories. Girls upto the age of 18 years and boys upto the age of 8 years would be allowed to stay in the Swadhar Greh with their mothers. (Boys of more than 8 years of age need to be shifted to the Children Homes run under JJ Act/ICPS.).

Implementing Agencies and Eligibility Criteria

Any of the following agencies/organizations can seek assistance under the Scheme:

- State Government agencies including Women Development Corporations established by the State Governments,
- Central or State autonomous bodies,
- Municipal Bodies,

- Cantonment Boards,
- Panchayati Raj Institutions and Co-operative institutions,
- Departments of Women and Child Development/Social Welfare of the State Governments which may construct Swadhar Greh and run them on their own or lease them to organization(s) having requisite experience for such period as deemed fit, for managing the operations under this scheme,
- Public Trusts registered under any law for the time being in force
- Civil Society Organizations (CSOs), such as NGOs etc. having proven track record of working in the fields of women's welfare/ social welfare/ women's education subject to the condition that such organization is registered under the Indian Societies Registration Act, 1860 or any relevant State Act.

The organization - Public Trusts and CSOs should meet the following criteria:

- It should be either recognized by the State/ UT Administration under existing Scheme/law or should be well known with the experience of working in the field for at least 3 years and its work should be reported to be satisfactory by the State Government/Union Territory Administration concerned,
- It should ordinarily have been engaged in the field of women's welfare/social welfare/women's education for a minimum period of two years prior to the request for grant-in-aid under the scheme,
- It should have facilities, resources, personnel and experience to undertake the management of such a project,
- Its financial position should be sound to take care of expenditure for a few months in case the grant is delayed,
- It should run the Swadhar Greh on a no – profit basis,
- The organization should have facilities like computers, internet connection etc. at Swadhar Greh.

Components of the Scheme

- Construction grant for construction of the building will be admissible to State Governments, Municipal Corporations,

Cantonment Boards and Panchayati Raj Institutions only. Land for this purpose is to be provided by the implementing agency free of cost. The grant shall be subject to a ceiling of Rs. 1,33,000/- per resident.

- Rent for Swadhar Greh, if run in a rented building. The maximum rent admissible for a Swadhar Greh intended for 30 residents is Rs. 50,000/- per month in grade 'A' cities, Rs. 30,000/- per month in grade 'B' cities and Rs. 18,000/- at other places.
- Assistance for recurring and non-recurring expenditure for management of Swadhar Greh.
- Provision for food, shelter, clothing, medical care, pocket expense for residents and children.
- Provision of counseling, legal aid, vocational training and guidance.

Support to Training and Employment Programme for Women (STEP)

The Ministry has been administering 'Support to Training and Employment Programme for Women (STEP) Scheme' since 1986-87 as a 'Central Sector Scheme'. The STEP Scheme aims to provide skills that give employability to women and to provide competencies and skill that enable women to become self-employed/entrepreneurs. The Scheme is intended to benefit women who are in the age group of 16 years and above across the country. The grant under the Scheme is given to an institution/ organisation including NGOs directly and not the States/ UTs. The assistance under STEP Scheme will be available in any sector for imparting skills related to employability and entrepreneurship, including but not limited to the Agriculture, Horticulture, Food Processing, Handlooms, Tailoring, Stitching, Embroidery, Zari etc, Handicrafts, Computer & IT enable services along with soft skills and skills for the work place, such as spoken English, Gems & Jewellery, Travel & Tourism, Hospitality.

Objectives of the scheme:

The scheme has 2 fold objectives viz:

- To provide skills that give employability to women.
- To provide competencies and skills that enable women to become self-employed/entrepreneurs.

Target groups: The scheme is intended to benefit women who are in the age group of 16 years and above.

Nari Shakti Puraskar

Ministry of Women and Child Development, has revised the guidelines for Women Awards for conferring on eminent women, organisations and institutions. These awards will be called "Nari Shakti Puruskars". Now, from the year 2016, 20 Nari Shakti Puruskars shall be conferred every year. The awards will be conferred on 8th March on the occasion of International Women's Day (IWD). The Award in each category shall carry a Certificate and a cash amount. The award would be given to eminent or outstanding Institutions or organizations and individuals.

The Ministry of Women and Child Development, in order to recognize the efforts made by women & institutions in rendering distinguished services for the cause of women, especially vulnerable and marginalized women, hereby announce national level awards for conferring on eminent women, organisations and institutions. These awards may be called 'Nari Shakti Puruskars'. The award shall be applicable to institutions and individuals who have rendered distinguished services to the cause of women.

Objectives:

1. In the last decade, there has been concerted effort by the Government to recognise and encourage women as reflected through a National Policy for Empowerment of Women in 2001. The issues related to women has gained utmost importance and focused attention. "Nari Shakti Puruskars" shall showcase the Government's commitment towards women with the aim of strengthening their legitimate place in the society. It will also provide an opportunity to the current generation to understand the contribution of women in building of society and the nation.
2. "Nari Shakti Puruskars" would be conferred on eminent women and institutions rendering distinguished service to the cause of women especially belonging to the vulnerable and marginalized sections of the society. The recipients would be drawn from institutions and individuals. The Ministry of Women & Child Development (herein after referred to as " the Ministry) would invite nominations from the State Governments, Union Territory

Administrations, concerned Central Ministries, Non-Governmental Organizations, Universities, Institutions, private and public sector undertakings (PSUs) working for empowerment of women.

Nari Shakti Puruskars shall be conferred every year. The recipients of the Puruskars shall be declared every year on 20th February and awards will be conferred on 8th March on the occasion of International Women's Day (IWD).

Indira Gandhi Matritva Sahyog Yojana (IGMSY) - A Conditional Maernity Benefit Scheme

Indira Gandhi Matritva Sahyog Yojana (IGMSY) is a Conditional Maternity Benefit (CMB) Scheme of the Government of India launched in 2010. The scheme is being implemented by the Ministry of Women and Child Development (WCD), Government of India. It is a conditional cash transfer scheme for pregnant and lactating mothers of 19 years of age or above for the first two live births. It provides partial wage compensation to women for wage-loss during the childbirth and childcare, and aims to promote conditions for safe delivery, good nutrition, and feeding practices. In 2013, the scheme was brought under the National Food Security Act (NFSA), 2013 to implement the provision of cash maternity benefit of Rs. 6,000 stated in the Act.

Objectives of IGMSY

i) Promoting appropriate practice, care and institutional service utilization during pregnancy, delivery and lactation;

ii) Encouraging the women to follow (optimal) nutrition and feeding practices, including early and exclusive breastfeeding for the first six months;

iii) Contributing to better enabling environment by providing cash incentives for improved health and nutrition to pregnant and lactating mothers.

Target Group: Pregnant Women of 19 years of age and above for first two live births (benefit for still births would be as per the guidelines of scheme). All Government/PSUs (Central & State) employees would be excluded from the scheme as they are entitled for paid maternity leave.

Mahila Police Volunteers (MPV)

MPV initiative is a voluntary service under which Mahila Police Volunteers (Gram Sakhi) will be inducted. It aims to bridge gap between Police and Civil society for safety of women. Under it, one MPV is envisaged per Gram Panchayat across the country to provide a link between police and community and facilitate women in distress. These MPVs will be selected through a laid out procedure from among the empowered, responsible, socially aware women who will facilitate police outreach on gender concerns. The volunteers will be trained by the respective state police authorities on their role and responsibilities. They will be appointed by state government.

Haryana became the first state in the country to launch Mahila Police Volunteer (MPV) initiative for safety of women. It was launched by Chief Minister Manohar Lal Khattar.

Mahila Police will serve as a public-police interface in order to fight crime against women. The broad mandate of MPVs is to report incidences of violence against women such as domestic violence, child marriage, dowry harassment and violence faced by women in public spaces. She will act as a role model for the community. MPVs are envisaged as empowered, responsible, socially aware women for fostering leadership in local settings to facilitate police outreach on gender concerns. They will be an interface between society and the police. Gender-Based Violence (GBV), faced by women both in public and private spaces, including domestic violence, sexual assault, rape, voyeurism, stalking. etc. is a major threat to women equality and empowerment. A gender-responsive police service requires specific training, the increased presence of female personnel and community outreach to integrate gender issues into policies, protocols and operational procedures. The Role of Police is pivotal in ensuring the safety and security of citizens in general and women in particular. To increase the visibility of women in the police force, Home Ministry has carried forward the initiative to give 33% reservation to women in the police force by implementing it in UTs and propagating in the States. Government of India promotes envisages engagement/ nomination of Mahila Police Volunteers (MPVs) in all States.

Benefits of Mahila Police Volunteers:

- MPV will serve as a public-police interface in order to fight crime against women

- MPVs is to report incidences of violence against women, such as domestic violence
- MPVs help to reduce crime, such as child marriage, dowry harassment, and violence faced by women in public spaces
- She will act as a role model for the community
- An MPV is an honorary position

Objectives of Mahila Police Volunteers:

1. An MPV will serve as a public-police interface in order to fight crime against women. The broad mandate of MPVs is to report incidences of violence against women such as domestic violence, child marriage, dowry harassment and violence faced by women in public spaces. She will act as a role model for the community.
2. An MPV is an honorary position.

Eligibility criteria for Mahila Police Volunteers:

1. Mahila police volunteer should be a female
2. She should not be less than 21 years of age
3. The incumbent must possess at least 12th pass certificate
4. She should be from the same geographical area and conversant with the local dialects
5. She must not have been convicted or imprisoned for the violation of any criminal law
6. No criminal proceedings must have been instituted against her
7. She should not be a member of any political party.

Mahila E-Haat

The Ministry of Women & Child Development launched "Mahila E-Haat" a bilingual portal on 7th March, 2016. This is a unique direct online marketing platform leveraging technology for supporting women entrepreneurs/SHGs/ NGOs for showcasing the products / services which are made/manufactured/ undertaken by them. It is an initiative for meeting aspirations and needs of women. This was done keeping in mind that technology is a critical component for business efficiency and to make it available to the majority of Indian women entrepreneurs / SHGs / NGOs.

The USP of this online marketing platform are:

- Facilitating direct contact between the vendors and buyers, as by displaying their contact number, address as also the basic cost of products/ services.
- Aadhar number is to be filled in mandatorily in the join us form, as a measure of id and for the payments to be received.
- For understanding how cash-less / digital transaction can be undertaken "Step by step instructions for various modes of payment: UPI, Wallets, PoS, and SMS banking (USSD)" is uploaded along with BHIM App information on the portal.
- Women need to be majorly involved in the value chain and have to be 18 years of age to display their products/ services. No illegal or contraband goods can be displayed. The vendors are allowed pricing their products and charging accordingly from the buyers.
- Being web based, it has unlimited reach and it is also very simple to join Mahila E-haat i.e. through the portal itself as the entire business of E-haat can be handled through a mobile. The vendor can be approached by the buyer physically, telephonically, by email, etc.

National Policies

The National Policy on Education-1986, revised in 1992(NPE) emphasises the need to use education as an agent of basic change in the status of women. The NPE proposes national education system to play a positive interventionist role in the empowerment of women, fostering of development of new values through redesigning of curriculum, text book, training and orientation of teachers, decision making and administrators and active involvement of educational institutions. These will be an act of faith and social engineering. Women studies will be promoted as the part of various courses and education institutions encouraged taking up active programs to further women's development.

Removal of women's illiteracy and obstacles inhibiting their access to, and retention in elementary education will receive overriding priority, through provision of special support services, setting up of time targets and effective monitoring. Major emphasis will be laid on women's participation vocational, technical and professional education at different levels. The policy of non-discrimination will be perused vigorously to eliminate sex, stereotyping in vocational and professional courses and promote women's participation in non-traditional occupations as well and existing and emergent technologies.

The National Policy of Empowerment of Women of 2001 has endorsed the provisions of NPE 1986. The policy prescribes:

- Equal access to education for women and girls.
- Special measures will be taken to eliminate discrimination.
- Universalize education.
- Eradicate illiteracy.
- Create a gender-sensitive educational system.
- Increase enrolment and retention rates of girls.
- Improve the quality of education.
- Development of occupation/vocation/technical skills by women.
- Reducing the gender gap in secondary and higher education.

Legal Framework- Women

- Commission of Sati Prevention Act, 1987 & Rules
- Dowry Prohibition Act, 1961
- Dowry Prohibition (Maintenance of Lists of Presents to the Bride & Bridegroom) Rules, 1985
- Indecent Representation of Women (Prohibition) Act, 1986
- National Commission for Women Act, 1990
- Protection of Women From Domestic Violence Act, 2005
- Sexual Harassment of Women at Workplace (Prohibition, Prevention & Redressal) Act, 2013 & Rules

Legal Framework- Children

- Child Marriage Restraint Act,1929
- Commissions for Protection of Child Rights Act, 2005

- Infant Milk Substitutes, Feeding Bottles & Infant Foods (Regulation of Production, Supply & Distribution) Act, 1992
- Infant Milk Substitutes, Feeding Bottles & Infant Foods (Regulation of Production, Supply & Distribution) Amendment Act, 2003
- Juvenile Justice (Care & Protection of Children) Act, 2000
- Juvenile Justice (Care & Protection of Children) Act, 2015
- Prohibition of Child Marriage Act, 2006
- Protection of Children from Sexual Offences Act, 2012

Legislation on Gender Based Issues

The following women specific Acts seek to address gender based violence and gender inequality:

- Dowry Prohibition Act, 1961:
- 1987: http://wcd.nic.in/commissionofsatiprevention.htm
- Indecent Representation of Women (Prohibition) Act, 1986
- Immoral Traffic Prevention Act, 1956
- Pre Conception and Pre Natal Diagnostic Technique Act, 1994 and The Pre-Natal Diagnostic Techniques (Regulation and Prevention of Misuse) Amendment Act, 2002
- Medical Termination of Pregnancy Act, 1971
- Maternity Benefit Act, 1961

Recent Important Acts pertaining to Gender Rights:

- The Protection of Women from Domestic Violence Act, 2005
- Hindu Succession (Amendment) Act, 2005
- The Prohibition of Child Marriage Act, 2006
- The Indian Divorce (Amendment) Act, 2001

Important Bills pertaining to Gender Rights

The following Bills have been drafted by various government bodies:

- Protection of Women against Sexual Harassment at Workplace Bill, 2010
- Sexual Offences Against Children Bill, 2010 (National Commission for Protection of Child Rights)
- The Assisted Reproductive Technologies Regulation Bill, 2010
- The Prevention of Communal and Targeted Violence Bill, 2011
- The Prevention of Crimes in the Name of Honour and Tradition Bill
- Recommendations and Suggestions on amendments to the Dowry Prohibition Act, 1961

Constitutional Provisions and Privileges

- Equality before law for women - ***Article 14***
- The State not to discriminate against any citizen on grounds only of religion, race, caste, sex, place of birth or any of them - ***Article 15*** (i)
- The State to make any special provision in favour of women and children - ***Article 15 (3)***
- Equality of opportunity for all citizens in matters relating to employment or appointment to any office under the State - ***Article 16***
- The State to direct its policy towards securing for men and women equally the right to an adequate means of livelihood - ***Article 39(a)***); and equal pay for equal work for both men and women - ***Article 39(d)***
- To promote justice, on a basis of equal opportunity and to provide free legal aid by suitable legislation or scheme or in any other way to ensure that opportunities for securing justice are not denied to any citizen by reason of economic or other disabilities -***Article 39 a***
- The State to make provision for securing just and humane conditions of work and for maternity relief - ***Article 42***

- The State to promote with special care the educational and economic interests of the weaker sections of the people and to protect them from social injustice and all forms of exploitation- *Article 46*
- The State to raise the level of nutrition and the standard of living of its people -*Article 47*
- To promote harmony and the spirit of common brotherhood amongst all the people of India and to renounce practices derogatory to the dignity of women - *Article 51(A) (e)*
- Not less than one-third (including the number of seats reserved for women belonging to the Scheduled Castes and the Scheduled Tribes) of the total number of seats to be filled by direct election in every Panchayat to be reserved for women and such seats to be allotted by rotation to different constituencies in a Panchayat - *Article 243 D(3)*
- Not less than one- third of the total number of offices of Chairpersons in the Panchayats at each level to be reserved for women - *Article 243 D (4)*
- Not less than one-third (including the number of seats reserved for women belonging to the Scheduled Castes and the Scheduled Tribes) of the total number of seats to be filled by direct election in every Municipality to be reserved for women and such seats to be allotted by rotation to different constituencies in a Municipality - *Article 243 T (3)*
- Reservation of offices of Chairpersons in Municipalities for the Scheduled Castes, the Scheduled Tribes and women in such manner as the legislature of a State may by law provide - *Article 243 T (4)*

Laws relating to working women

- Contract Labour (Regulation and Abolition) Act, 1976
- Employees State Insurance Act, 1948
- Equal Remuneration Act, 1976
- Factories (Amendment) Act, 1948
- Maternity Benefit Act, 1961 (Amended in 1995)
- Plantation Labour Act, 1951

Laws relating to marriage & divorce

- Anand Marriage Act, 1909
- Arya Marriage Validation Act, 1937
- Births, Deaths & Marriages Registration Act, 1886
- Bangalore Marriages Validating Act, 1936
- Converts' Marriage Dissolution Act, 1866
- Dissolution of Muslim Marriages Act, 1939
- Family Courts Act, 1984
- Foreign Marriage Act, 1969
- Hindu Marriage Act, 1955
- Hindu Marriages (Validation of Proceedings) Act, 1960
- Indian Christian Marriage Act, 1872
- Indian Divorce Act, 1869
- Indian Divorce Amendment Bill, 2001
- Indian Matrimonial Causes (War Marriages) Act, 1948
- Marriage Laws (Amendment) Act, 2001
- Marriages Validation Act, 1892
- Muslim Women (Protection of Rights on Divorce) Act, 1986
- Parsi Marriage & Divorce Act, 1936
- Prohibition of Child Marriage Act, 2006
- Special Marriages Act, 1954

Laws relating to maintenance

- The Code of Criminal Procedure, 1973
- Order for maintenance of wives, children and parents under section 125
- Procedure to be followed under section 125
- Alteration in allowance under section 125
- Enforcement of the order of maintenance

Laws relating to abortion

- Medical Termination of Pregnancy Act, 1971
- Pre-Natal Diagnostic Techniques
- Pre-Natal Diagnostic Techniques (Regulation & Prevention of Misuse) Amendment Act, 2001
- Pre-Natal Diagnostic Techniques (Regulation & Prevention of Misuse) Amendment Act, 2002

Laws relating to property, succession, inheritance, guardianship & adoption

- Guardians & Wards Act, 1890
- Hindu Adoptions & Maintenance Act, 1956
- Hindu Inheritance (Removal of Disabilities) Act, 1928
- Hindu Minority & Guardianship Act, 1956
- Hindu Succession Act, 1956
- Hindu Succession (Amendment) Act, 2005
- Indian Succession Act, 1925
- Indian Succession (Amendment) Act, 2002
- Married Women's Property Act, 1874
- Married Women's Property (Extension) Act, 1959

Offences against women and children in the Indian Penal Code

- The Indian Penal Code, 1860 (http://www.vakilno1.com/bareacts/indianpenalcode/indianpenalcode.htm)
- Abandoning of child under 12 years of age
- Adultery (http://www.vakilno1.com/bareacts/indianpenalcode/S497.htm)
- Assault or criminal force to a woman with intent to outrage her modesty
- Buying minor for purpose of prostitution (http://www.vakilno1.com/bareacts/indianpenalcode/S373.htm)
- Causing death of quick unborn child by act amounting to culpable homicide

- Causing miscarriage or miscarriage without the woman's consent
- Cohabitation caused by a man deceitfully inducing a belief of lawful marriage
- (http://www.vakilno1.com/bareacts/indianpenalcode/S493.htm)
- Concealment of birth by secret disposal of dead body
- Concealment of former marriage
- Death caused by act done with intent to cause miscarriage
- Dowry death
- Enticing, detaining or taking away with criminal intent a married woman
- Fraudulent marriage ceremony without lawful marriage
- Husband or relative of a husband of a woman subjecting her to cruelty
- (http://www.vakilno1.com/bareacts/indianpenalcode/S498A.htm)
- Importation of girl from foreign country
- Intercourse by man with his wife during separation
- Intercourse by a member of management or staff of a hospital with any woman in that hospital
- Intercourse by public servant with a woman in his custody
- Intercourse by superintendent of jail, remand home, etc.
- Kidnapping, abducting or inducing woman to compel her marriage
- Marriage ceremony fraudulently gone through without lawful marriage
- (http://www.vakilno1.com/bareacts/indianpenalcode/S496.htm)
- Marrying again during lifetime of spouse (http://www.vakilno1.com/bareacts/indianpenalcode/S494.htm)

- (http://www.vakilno1.com/bareacts/indianpenalcode/S495.htm))
- Preventing a child from being born alive or causing its death after birth
- Procreation of minor girl (http://www.vakilno1.com/bareacts/indianpenalcode/S366A.htm)
- Rape (http://www.vakilno1.com/bareacts/indianpenalcode/S375.htm)
- Selling minor for purpose of prostitution
- Word, gesture or act intended to insult the modesty of a woman.

Digital Gender Atlas for Advancing Girls' education

The Ministry of Human Resource Development, Department of School Education and Literacy.

The tool, which has been developed with the support of UNICEF, will help identify low performing geographic pockets for girls, particularly from marginalised groups, such as scheduled castes, schedule tribes and Muslim minorities, on specific gender related education indicators.

Objectives

- To identify low performing geographic pockets for girls,particularly from marginalised groups, such as scheduled castes, scheduled tribes and Muslim minorities, on specific gender related education indicators.
- To ensure equitable education with a focus on vulnerable girls, including girls with disabilities, the Gender Atlas has been developed as a hands on management tool to enable critical decisions and action in pockets where performances are below par.

Salient features

- The architecture of the Atlas enables dynamic navigation between geographical representation and numeric data that presents State, District and Block level status on key parameters for girls' education at primary, upper primary and secondary levels. Thus, upon making a change in either the Search/Geo Search Tabs, the maps on the screen automatically change.

- The Atlas provides a comparative composite index based on quartile ranking of gender related indicators at National, State, District and Block levels.
- The Atlas enables a trend analysis and tracking of performance of individual gender related parameters across periods of time.
- The Atlas is constructed on an upon source platform with an in-built scope of updating data by authorized persons to retain its dynamic character.

Components of the Digital Gender Atlas

The main components of the Gender Atlas are :

- Composite Gender Ranking : Selected gender related indicators have been identified for primary, upper primary and secondary education. A composite gender index has been developed using the formulae that are used for calculating the EDI in D1SE. Accordingly, the selected gender indicators have been grouped into four-Access, Infrastructure, Teachers and Outcomes.

  The Atlas provides an aggregated picture of performance through a comparative visualization of composite gender index based quartile ranking of gender related indicators at National, State, District and Block levels.
- **Trend Analysis of Gender Indicators** - The Atlas enables a trend analysis and tracking of performance of individual gender related indicators (that have been used for the composite gender ranking) across a period of three years.
- Vulnerabilities based on educational indicators in districts with substantial tribal, schedule caste, minority population in educationally backward blocks and in left wing extremist districts, and the low sex ratio districts selected under Beti Bachao, Beti Padhao programme.
- **Spatial Display of Pockets with Multiple Disadvantages Related to Education:** Spatial identification of pockets in states or blocks in districts where juxtaposition of key relevant gender indicators reveal maximum disadvantages, viz. education development.

- **Vulnerabilities:** This component of the Atlas visualizes the vulnerability status of an area based on a composite index of three broad indicators.
  - Rural female literacy - from Census, 2011
  - Percentage girls/boys married below the legal age of marriage-DLHS and working children (bifurcated by gender and age) groups of 5-14 years and 15-19 years. Census of India, 2011. These maps are available upto the district level.
- **Children with Disabilities:** This part of the Atlas visualizes the performance status of a single indicators related to children with disabilities, particularly girls, based on quartile ranking across three years i.e. 2011-12, 2012-13 and 2013-14. The performance status of an individual indicator can be visualized at state the district level.

(*Source:* Digital Gender Atlas for Advancing Girls' Education)

❑❑❑

Chapter 15

Crisp Notes on Gender

Gender Analysis: It is a tool for examining the differences between roles that women and men play, the different levels of power they hold, their differing needs, constraints & opportunities and the impact of these differences on their lives.

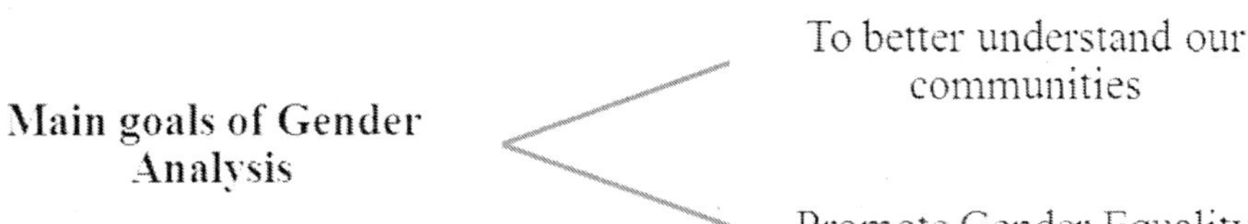

When should Gender Analysis be used?

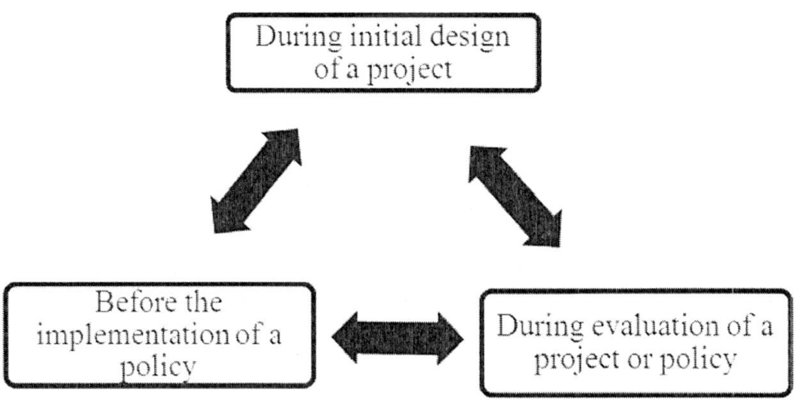

Methods of Gender Analysis

Different methods of Gender Analysis

Traditional Method	Non-Traditional
• Formal Interviews • Survey • Mapping • Research through libraries and organisation	• Household Interviews • Focus group sessions • Informal conversations • Walking Tours • Observing community practices

Gender Analysis Framework

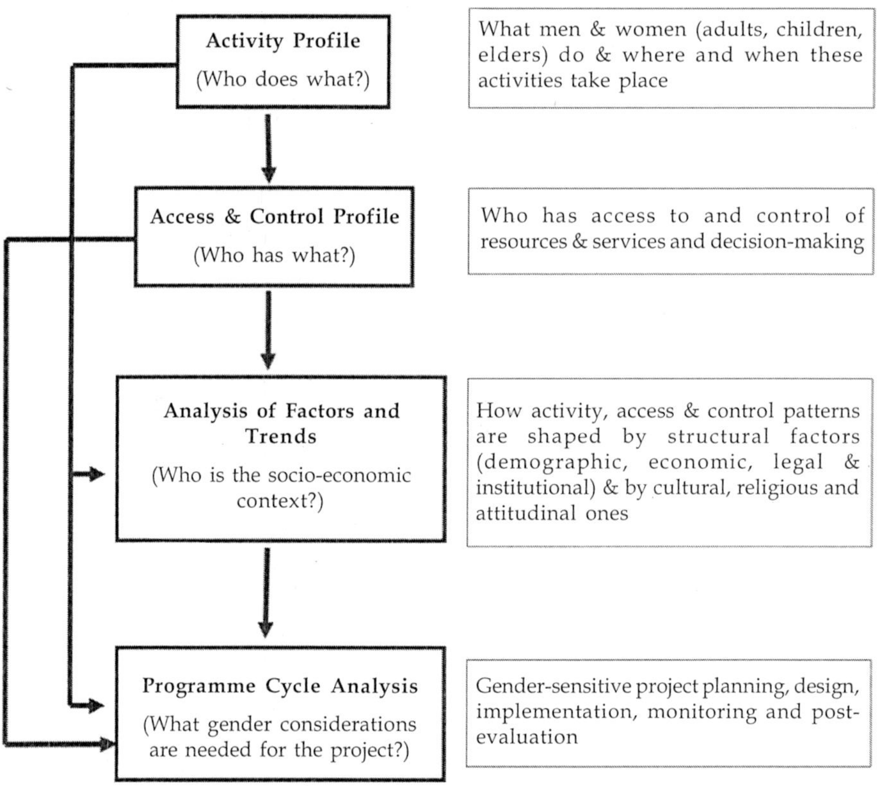

Types of Gender Analysis

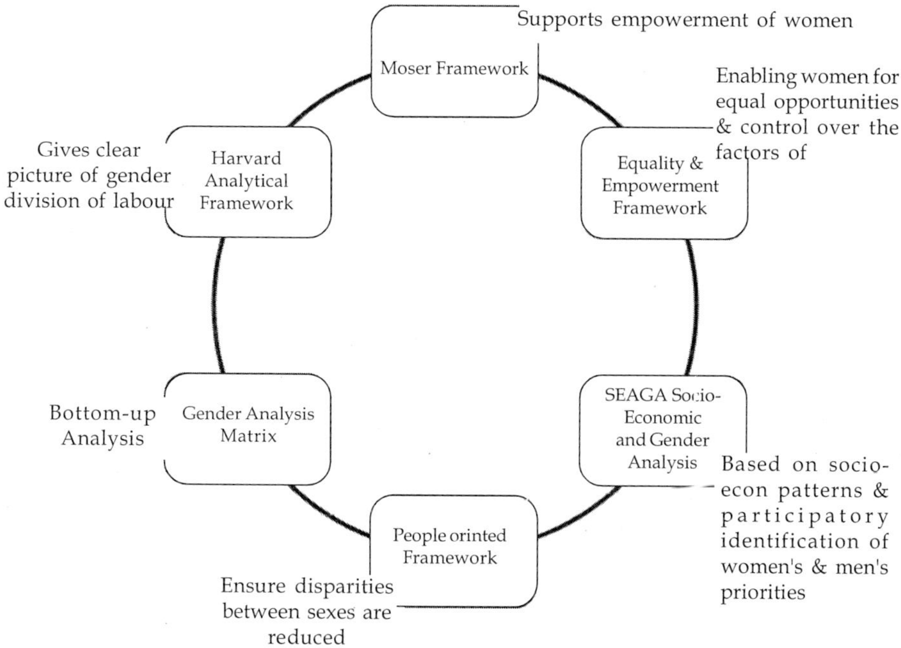

Tool for Gender Analysis

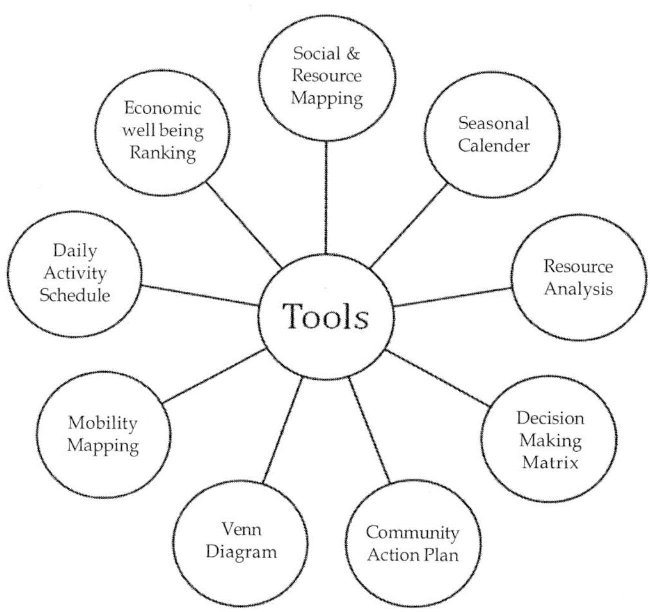

Gender Analysis Matrix

Def: It is an analytical tool that uses participatory methodology to facilitate the definition and analysis of gender issues by the community that are affected by them.

Aims:

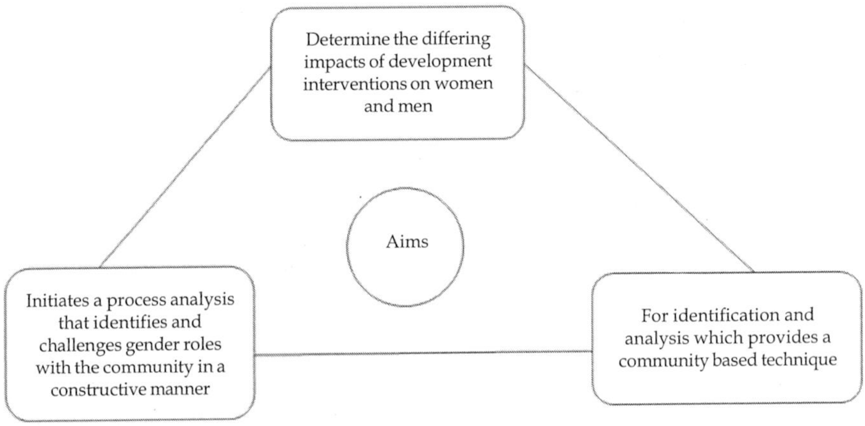

Principles :

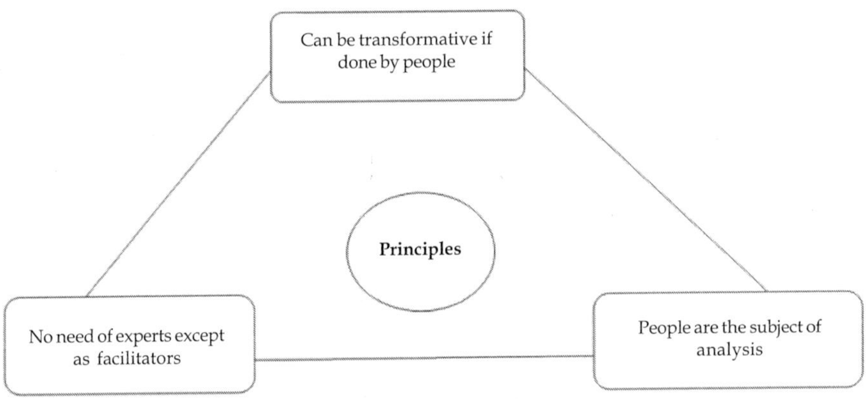

What: A tool for gender analysis of development (programme / project) at community level

Why: to determine differing impacts of intervention on women and men

Who: done by a group within the community which includes men and women in equal number

When: Planning stage, design stage and monitoring stage

Level of Analysis:

- **Women** - All ages in the community
- **Men** - All ages in the community
- **Household** - All men, women, and children residing together
- **Community** - Everyone in the project area as a whole

Category of Analysis:

- **Labour** - change in task, capacity
- **Time** - change in the amount of time
- **Resource** - Change in access to capital
  Change in control over resources
- **Cultural** - Change in social impact

Format

Particulars	Labour	Time	Resources	Culture
Women				
Men				
Household				
Community				

Differences between Gender Equality and Gender Equity

Gender Equality	Gender Equity
Gender equality requires equal enjoyment by women and men of socially-valued goods, opport-unities, resources and rewards	Gender equity is the process of being fair to women and men.
Gender equality includes all aspects of gender equity	Equity leads to equality
'Gender equality' is the term used in human rights treaties, such as The Convention on the Elimination of all forms of Discrimination Against Women (CEDAW).	The term was first proposed in 1995 for the Beijing Platform for Action by Islamic fundamentalists and Vatican followers who sought to allow parents to divide their

	property between their children according to their respective 'needs' in their gender-specific roles.
Absence of discrimination based on sex	Equal treatment before the law
It is based on premise that women and men should be treated in same way	Takes into consideration the differences in women's and men's lives
Fails to recognize that equal treatment will not produce equitable results because women and men have different life experiences	Recognizes that different approach may be needed to produce outcomes that are equitable

Why Gender equity is important to achieve Gender equality?

Women and men should not only be given equal access to resources and equal opportunities, but they should also be given the means of benefiting from this equality. This is where the concept of 'gender equity' comes into play. Gender equity implies fairness in the way women and men are treated. The different life experiences and needs of men and women are taken into consideration and compensation is made for women's historical and social disadvantages. The lower status of women in society often constitutes a handicap and provisions should be made to redress this inequality before they can take advantage of the opportunities provided. Gender equity, thus, serves to level the playing field and empower women. Therefore, we can say that equity is essential to achieve true equality.

Why Gender equality is more important than equity?

Gender equality is preferred term rather than gender equity. Gender equity denotes an element of interpretation of social justice, usually based on tradition, custom, religion or culture which is most often to the detriment to women. Such case of equity in relation to the advancement of women is unacceptable.

Gender equality means that the rights, responsibilities and opportunities of individuals will not depend on whether they are born male or female. Equality doesn't mean "the same as" - promotion of gender equality doesn't mean the women and men will be same.

Equality between women and men has both a quantitative and a qualitative aspect. The quantitative aspect refers to the desire to achieve representation. Equitable representation of women is increasing balance and parity, while the quantitative aspect refers to achieving equitable influencing on establishing development priorities and outcomes for women and men.

Equality involves ensuring that the perceptions, interests, needs and priorities of women and men will be given equal weight in planting and decision making.

For sustainable people-centered development, equality is more important.

Distinctions and controversy over terms Gender equity

This refers to the treatment of people according to their respective needs to ensure equal rights, obligations and opportunities. It recognises that people are born with different strengths and weaknesses and demands that development practitioners adopt different actions accordingly. A limitation to this term is society's subjective comprehension of what people's 'needs' are and what action should be taken towards equity. The term was first proposed in 1995 for the Beijing Platform for Action by Islamic fundamentalists and Vatican followers who sought to allow parents to divide their property between their children according to their respective 'needs' in their gender-specific roles. In this way 'gender equity' is a problematic term as it may inspire legislation that further encourages a person to fill a predetermined gender-defined role, and perpetuate gender inequality. Consequently, it is necessary to focus the outcome goal on equity toward equality of opportunity.

Gender equity is a mechanism through which gender equality may be achieved. A consideration of both gender equity and gender equality is integral to good development practice. The aim of a development practitioner should be equity for equality.

Gender Sensitization

Def: Recognition of the differences and inequities between women's and men's needs roles, responsibilities and identities.

Gender Sensitization Process

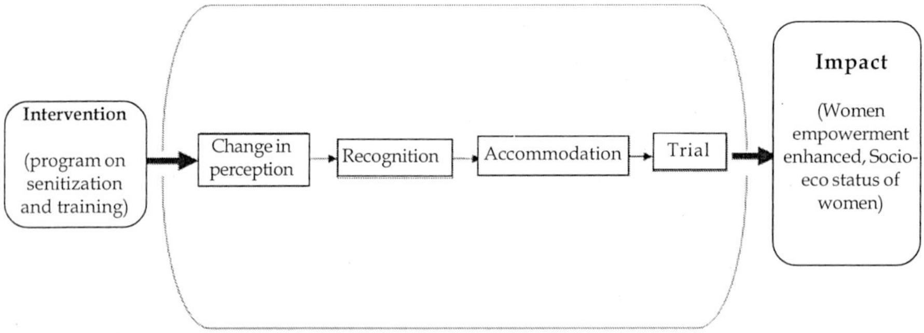

Gender Mainstreaming

Def: The process of assessing the implications for women and men of any planned action in all areas and at all levels. (U.N)

Ultimate goal of Gender mainstreaming is Equality.

Strategy for Gender Mainstreaming

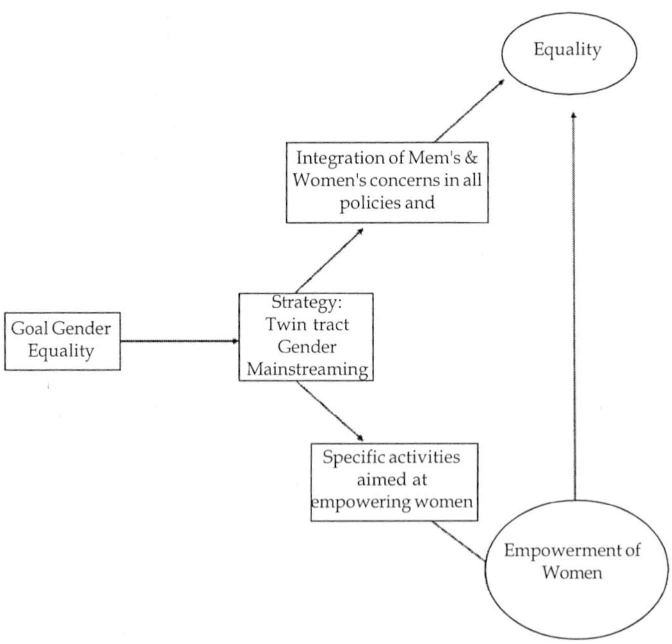

Requirements for Gender Mainstreaming

1. Strategy
2. Relevance
3. Hands on technical Assistance
4. Skills and expertise in gender
5. Financial resources
6. Accountability Mechanisms that use gender specific
7. Leadership

Needs for Gender Mainstreaming

1. To support promotion of gender equality at all levels
2. To participate in and equally benefit from development effort
3. Benefit to both beneficiaries and project implementers and stakeholders
4. Necessary measures designed to combat the direct and indirect consequences of past discrimination
5. To complement with targeted interventions to promote gender equality and women empowerment

Tools for Gender mainstreaming

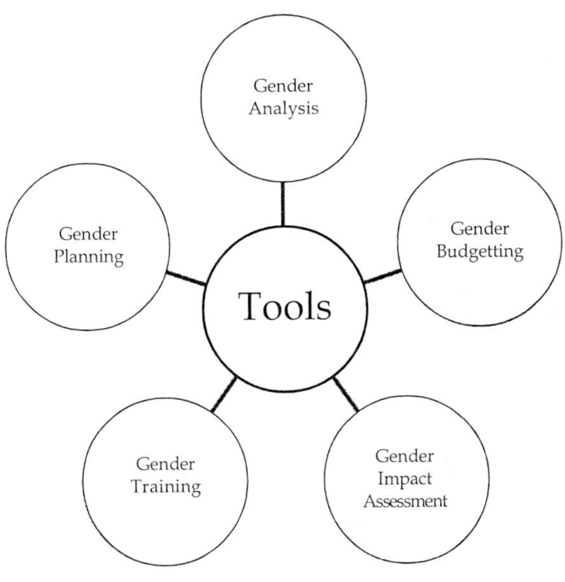

Gender mainstreaming in the program cycle

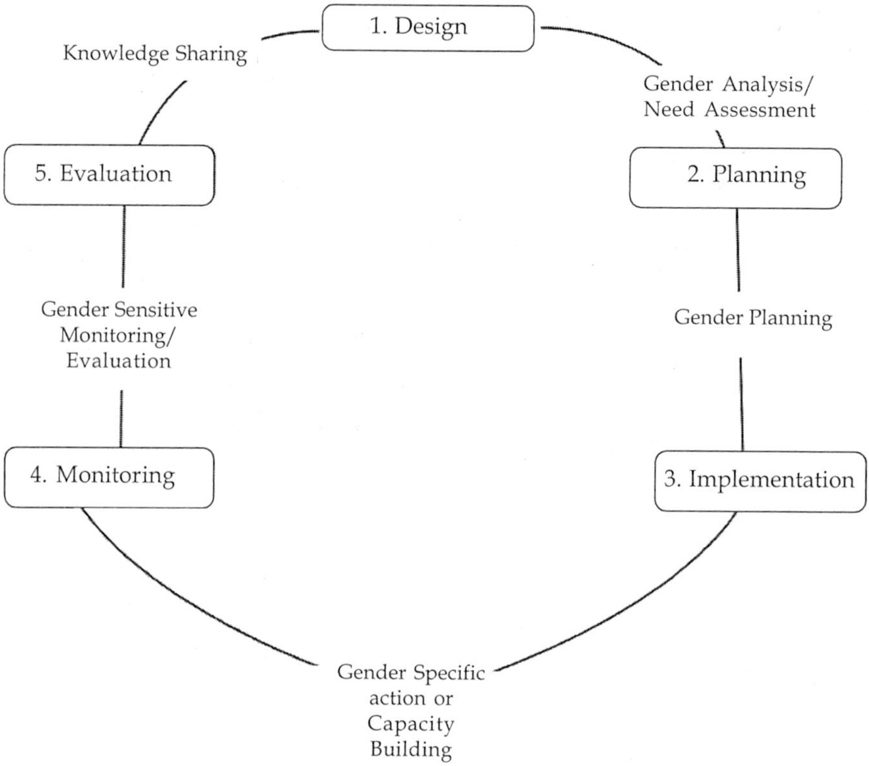

Gender Budgeting

Gender budgeting is a part of the gender mainstreaming strategy. It entails a gender based assessment of budgets incorporating a gender perspective at all levels of the budgetary process & restructuring revenues and expenditures in order to pomote gender equality.

Gender budgeting focuses on a gender based analysis and an equality-oriented evaluation of the distribution of resources. These resources are mainly money, time as well as paid and/or unpaid work. Gender budgeting is a process that entails incorporating a gender perspective at all levels and stages of budgetary process.

Gender budgeting in India started from 2005-06 (Eleventh five-year plan)

Gender Budgeting in India - Evaluation

- **Seventh Plan (1987-1992):** 27 major women specific schemes identified for monitoring to assess quantum of funds/benefits flowing to women.
- **Eight Plan :** Highlighted need to ensure a definite flow of funds from general sectors to women.
- **Ninth Plan :** Women's component Plan (WCP) - 30% of funds were sought to be ear-marked in all women related sectors.
- **Tenth Plan :** Aims at tying up concepts of WCP & Gender Budgeting.
- **Eleventh Plan :** States that gender equity requires adequate provisions to be made in police and schemes across ministried and departments. It also entails stict adhence to gender budgeting across the board.

Components of gender budget

Gender budgeting in India follows the following measures in accounting gender specific programmes:

i) Specifically targeted expenditure to women and girls;

ii) Pro-women allocations, which are the composite expenditure schemes with a significant women's component; and

iii) Residual public expenditures that have gender–differential impacts.

Aims of Gender Budgeting

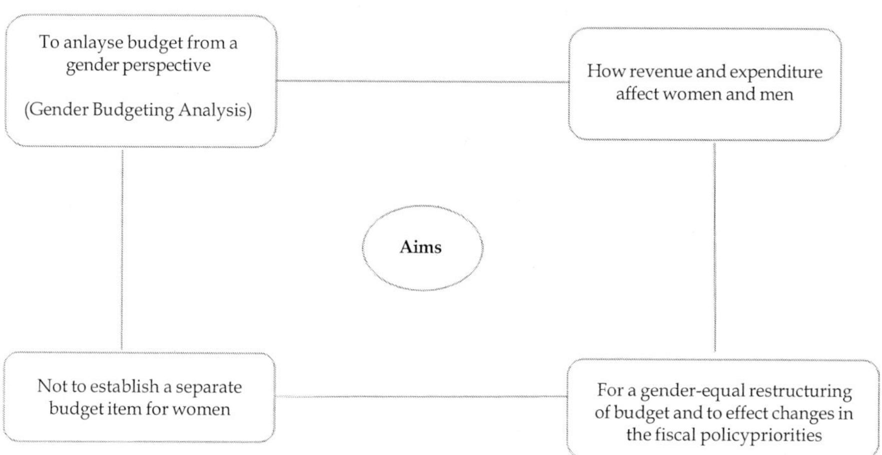

Five-step framework of Gender-Budgeting

Analysis of the situation of women and men in a given sector

↓

Assess whether the policy addresses the gender issues and gaps

↓

Assess the adequacy of budget allocations for implementation

↓

Monitoring whether the money is spent as per planned

↓

Impact assessment of policy / program

Gender Budgeting - From policy to outcomes

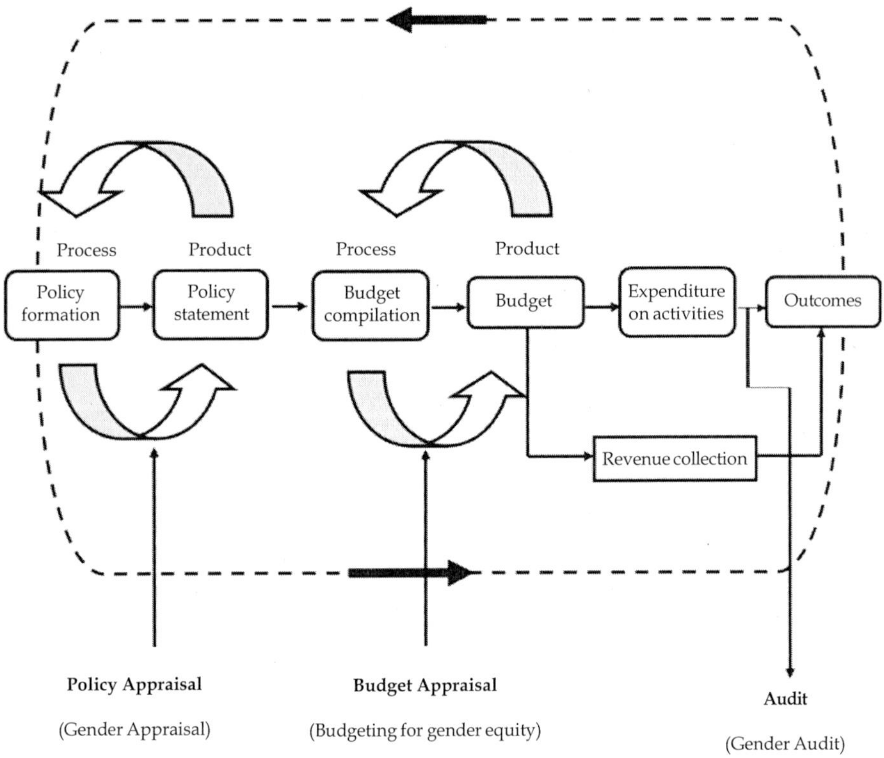

Model for Empowerment of Women through SHG

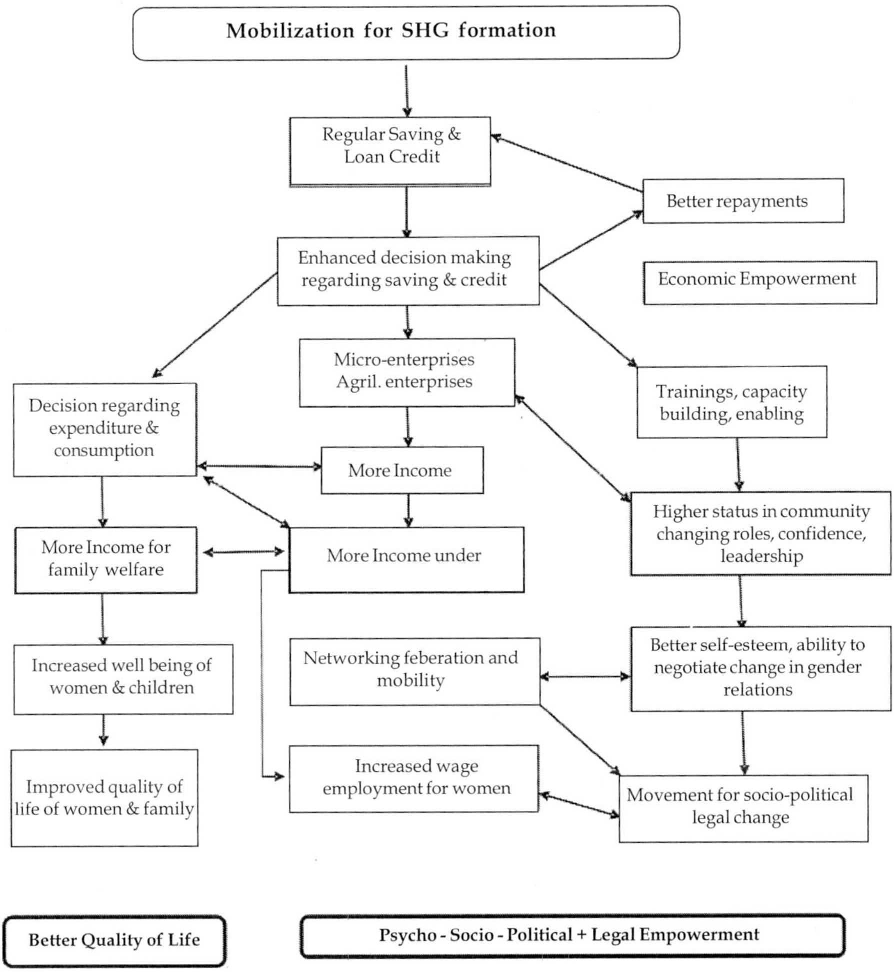

- Enabling + Motivation = Empowerment
- SHG-Synergy + Group Efficacy
- Power within + Power with + Power to

} Empowerment of SHG women

Women Empowerment Analytical Framework

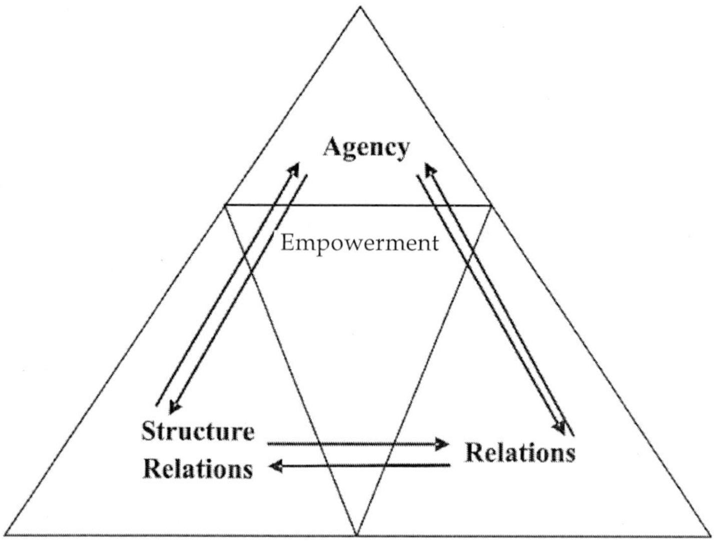

Core Sub-Dimensions of Empowerment	Agency - Level Indicators	Structural Level Indicators	Relation Indicators
Notion of Self-worth; Dignity Control & Influence over Household & Public resources Policy Integrity Collective Effort Solidarity	• Knowledge of rights & structures of gender inequality • Education Level • Changes in Self -Image	• Participation in political process • Equitable access to basic human resources • Legal change &/or enforcement of women • Control of strategic resources • Pro-women changes in family /kin norms & Institutions • Equal economic opportunity including	• Participation in society / solidarity groups; connection with other groups • Influence on formal & informal decision-makers • Influence on male attitudes regard to gender roles & norms.

	ownership/ Control of strategic assets • Pro-women state budgets & Develop- ment Policies.

'Women Empowerment' concept was introduced at the International Women's Conference at Nairobi 1985.

Def of Empowerment: The process of enhancing the capacity of individuals and groups to make choices and to transform those choices into desired action and outcomes. (World Bank, 1997).

Gender Empowerment Measure: whether women and men are able to actively participate in economic and political life and take part in decision-making.

Women Empowerment has five components

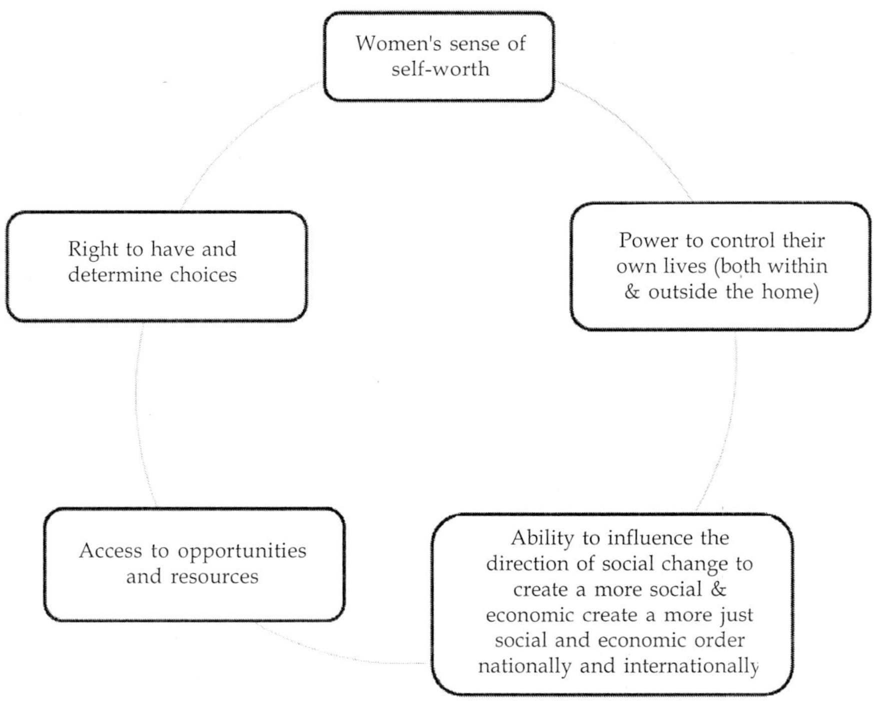

Sustainable Livelihoods through a Gender Lens

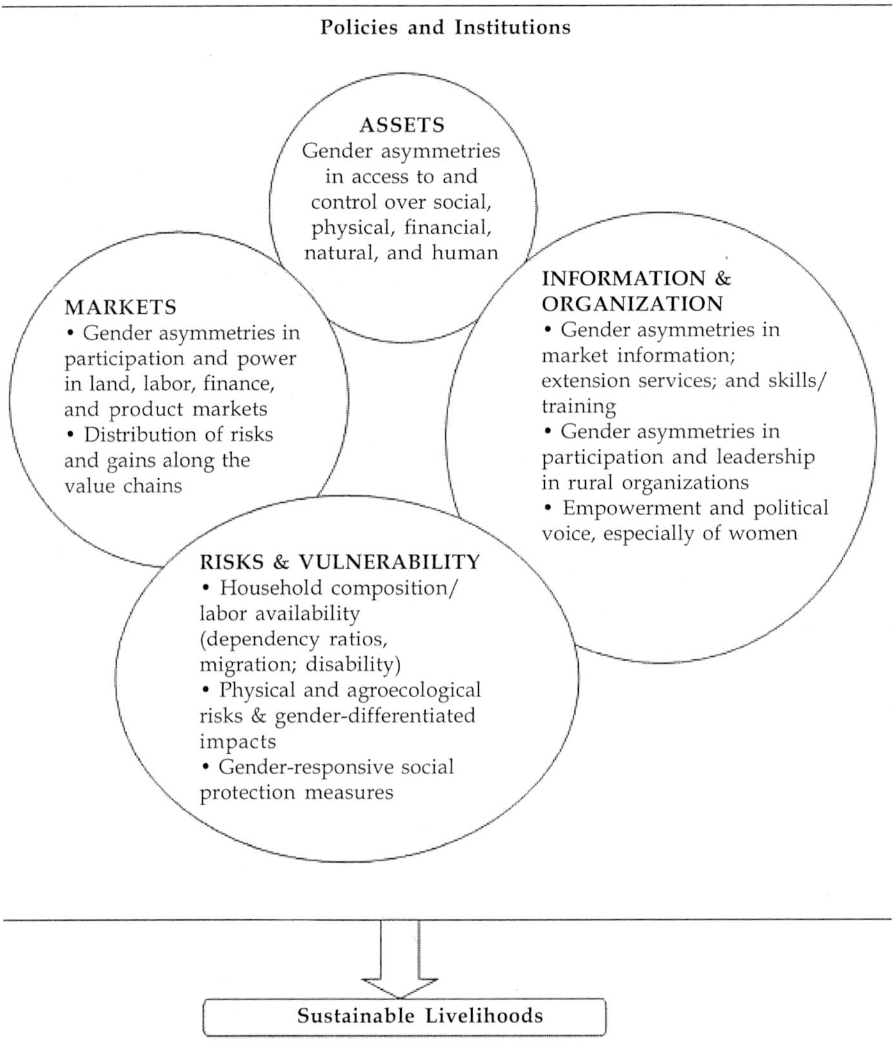

Five-dimensional framework for gender inequalities in agriculture

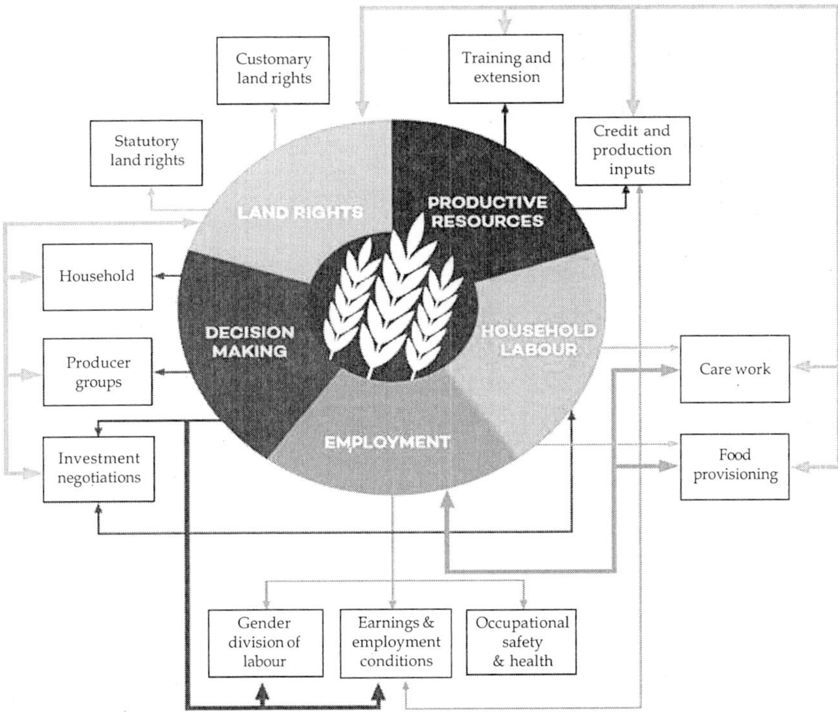

Source: Sexsmith *et al.* (2017)

❑❑❑

Easy Bytes for a Quick Reference

India's First Woman President	Pratiba Patil
India's First Woman Prime Minister	Indira Gandhi
India's First Woman Governor	Sarojini Naidu
India's First Woman IPS Officer	Kiran Bedi
First Woman Chief Minister of State	Sucheta Kripalani
First Woman Union Minister	Rajkumari Amrita Kaur
First Woman President of INC	Annie Besant
First Woman Judge of Supreme Court	Meera Sahib Fatima Bibi
First Woman to get Ashok Chakra	Nirja Bhanot
First Indian Woman Ambassador at UN	Vijaylakshmi Pandit
First Indian Woman to get Nobel Prize	Mother Terasa
First Indian Woman to swim across English Channel	Arati Saha
First Indian Woman to climb Mt. Everest	Bachendri Pal
First Indian Woman to become Miss World	Miss Reita Faria
First Indian Woman to climb Mt. Everest twice	Santosh Yadhav
First Indian Woman to become Miss Universe	Sushmita Sen
First Indian Woman to get Bharat Ratna	Indira Gandhi
First Indian Woman to get Jnanpith Award	Ashapurna Devi
First Indian Woman to win WTA Title	Sania Mirza

First Indian Woman Airline Pilot	Durga Banerjee
First Indian Woman to win Gold in Asian Games	Kamaljeet Sandhu
First Indian Woman President of INC	Sarojini Naidu
First Indian Woman to Booker Prize	Arundhati Roy
First Woman Musician to get Bharat Ratna	MS Subbalakshmi
First Indian Woman to go to space (2003)	Kalpana Chawla
First woman pilot	Sarla Thakral
First Woman Mountaineer to Climb Everest (1984)	Bachendri Pal
First Photo Journalist (1913-2012)	Homai Vyarawalla

□□□

References

Adhiguru P., Birthal P S and Kumara B G. (2009). Strengthening Pluralistic Agricultural Information Delivery Systems in India. Agricultural Economics Research Review **22**: 71–79.

Ahmad N. (2011). From Policy to Projects: Rapid Gender Review/Screening Tool, MDB-Sponsored Regional Workshops to Mainstream Gender Equality in Infrastructure Policies & Projects Africa Regional Workshop. Addis Ababa: Social Development Department, The World Bank.

Akudugu M A, Guo E, and Dadzie S K. (2012). Adoption of Modern Agricultural Production Technologies by Farm Households in Ghana: What Factors Influence Their Decisions? Journal of Biology, Agriculture and Healthcare **2**(3):1–13.

Alkire S, Meinzen-Dick R, Peterman A, Quisumbing A, Seymour G, and Vaz A. (2013). The Women's Empowerment in Agriculture Index.World Development, **52**: 71–91.

Annual Report. (1990-91). NABARD, p. 19.

Annual Report. (1998). NABARD, p. 1.

Annual Report. (2010). Office of DRDA, DUDA and MAVIM, Amravati.

Anonymous (2010-11). Report of Food and Agriculture Organization on women in agriculture: Closing the gender gap for development.

Anonymous (2012). India ranked at 115 on women empowerment scale: Survey. The Economic Times, October 16th, 2012.

Anonymous. (1996). Women, Gender and Development, Programme Handbook, Royal Tropical Institute, The Netherlands.

Anonymous. (2007). Tamil Nadu Corporation for Development of Women Ltd. - Credit guidelines for SHGs, Handbook, 2007, p. 5.

Aruna R, Anderson M K and Overholt C. (1991). Gender Analysis in Development Planning: A Case Book. West Hartford, C T: Kumarian Press.

Asian Development Bank. (2013). Gender Equality and Food Security—Women's Empowerment as a Tool against Hunger. Mandaluyong City: Asian Development Bank.

Bacon C. (2010). A spot of coffee in crisis: Nicaraguan smallholder cooperatives, fair trade networks, and gendered empowerment. Latin American Perspectives **37**(2): 50–71.

Bannerjee B N. (2001). My Great India: Economic Development & Glaring Disparities. New Delhi: CBS Publishers & Distributors.

Belurkar G M., Wakle P. K. & Gholve M. A. (2003). A study on decision making pattern and participation of rural women in animal husbandry and dairying enterprise. *Maharashtra Journal of Extension Education,* **22** (2): 81-85.

Bennett L. (1993). Women, poverty and productivity in India. Economic Development Institute of the World Bank.

Biogas Co-operative . Togetherness is the key, Women, Energy and Development, Urja Bharati June 2001, Ministry of Non-Conventional Energy Sources, New Delhi

Case studies on programmes: "Planning and Management", Self Help Group and Microfinance, p. 49-52.

Census (2001). Census Report, Office of the Registrar General and Census Commissioner, Ministry of Home Affairs, Government of India.

Census (2011). Census Report, Office of the Registrar General and Census Commissioner, Ministry of Home Affairs, Government of India.

Chacharkar D Y and Satpute R C (2008). "New emerging cooperatives - A critical study of SHG", NCDC Bulletin, Dec., p. 2-3.

Chandrakavate M S. (2006). "The SHG model of microfinance : A salient movement towards empowering rural women", *Southern Economist,* **44**(17): 31-32.

Chaturvedi S V. (2015). Participatory Gender Auditing: As A Process For Gender Mainstreaming. *Journal of Management Engineering and Information Technology,* **2**(1):2-6.

Chayal K., Daaka B L and Suwalka R L (2009). Analysis of role performed by farm women in dairy farming. *Indian Journal of Dairy Science,* **62**: 491-494.

Coomaraswamy R (1996). Preliminary Report of the UN Special Rapporteur on violence against women , 1996, p 49.

Dankelman I and Davidson J (2013) .Women and the Environment in the Third World: Alliance for the Future. London: Earth scan.

European Commission. (2004). EQUAL Guide on Gender Mainstreaming. Available at: http://ec.europa.eu/employment_social/equal/data/document/gendermain_en.pdf

Farnworth C and Hutchings J (2009). Organic agriculture and women's empowerment. Germany: IFOAM. Retrieved from https://grist.files.wordpress.com/2011/02/gender-study-090421.pdf

Financing a Self Help Groups - NABARD, 1998, p. 28.

Fletschner D and Kenney L (2011). Rural Women's Access to Financial Services Credit, Savings and Insurance, ESA Working Paper No. 11. Rome, Italy: Agricul-tural Development Economics Division, The Food and Agriculture Organisation of the United Nations.

Flintan F. (2008). Women's Empowerment in Pastoral Socities. World Initiative for Sustainable Pastoralism, GEF, UNDP & International Union for Conservation of Nature: 1-125.

Fulzele R M and Meena B L (1995). Training needs of dairy tribal women. *Indian Journal of Dairy Science*, **48**: 551-553.

Gender Budgeting Handbook for Govt. of India. (2007). Ministries and Departments, Ministry of Women and Child Development, Govt. of India.

Gender Equality and Food Security, Published by FAO.

Gender Reference Guide http://www.snvworld.org/gender/gender-mainstreaming_analysis_1

George P S and Chokshi (1990). Dairy development decisions at farm level. Monograph No.66, IIM, Ahmedabad.

Gite L P, Singh S P, Agarwal N, and Majumder J. (2007). Women Friendly Improved farm Tools and Equipment. Technical Bulletin No. CIAE/2007/128. Bhopal, India: Central Institute of Agricultural Engineering.

Government of India. (2007). Gender Budgeting Hand Book for Government of India Ministries & Departments. Ministry of Women and Child Development.

Government of India. (2015). National Gender Resource Centre in Agriculture. Ministry of Agriculture and Farmers Welfare, New Delhi

Grover I. (2003). Unit-I Gender Theory and Development: Post Graduate Diploma on Gender and Development through Distance Education, HEL Project, Department of HSEE, CCS HAU, Hisar, India

Guidelines for Gender analysis of Public Expenditure, Ministry of Women and Child Development, Govt. of India.

Hanson L, Terstappen V, Bacon C M, Leung J, Cuenca A, Flores D A, and Rojas M A. (2012). Gender, health, and Fairtrade: Insights from a research-action programme in Nicaragua. Development in Practice, **22**(2):164–179.

Harriss-White B and Gooptu N. (2001). Mapping India's World of Unorgan-ised Labour, Socialist Register. **37**: 89– 118.

ICAR-Central Institute for Women in Agriculture (CIWA). 2016. Annual Report 2015–16. Bhubaneswar, India, p.10.

John C R & Thirunavukkarasu M. (2002). Role of farm women in livestock keeping. *Indian Dairyman*, **54**: 35-40.

Johnson J N., Govindaradjane S and Sundararajan T (2013). The role of rural women in watershed development project. *International Journal of Engineering Science and Innovative Technology*, **2** (1): 540-544.

Kamalakannan K K (2005). "The role of financial institution in development of women entrepreneur", *Kurukshetra*, **53**(16):12.

Karmakar K G (2003). "Rural credit and self help groups - Microfinance needs and concepts in India", New Delhi, *Sage Publication India Pvt. Ltd.*, p. 231.

Kasente D (2012). Fair Trade and organic certification in value chains: Lessons from a gender analysis from coffee exporting in Uganda. Gender and Development, **20**(1): 111–127.

Kelkar G (2011). Gender and Productive Assets: Implications for Women's Economic Security and Productivity. Economic & Political Weekly **46**(23):59–68.

Kishor S & Gupta K (2009). Gender equality and women's empowerment in India. National Family Health Survey India. Ministry of Health and Family Welfare, Government of India, New Delhi.

KIT, Agri-ProFocus and IIRR. (2012). Challenging chains to change: Gender equity in agricultural value chain development. Amsterdam: KIT

Publishers, Royal Tropical Institute. Retrieved from https://www.cordaid.org/en/wpcontent/uploads/sites/3/2013/02/Challenging_chains_to_change.pdf

Lahoti S R, Chole S R and Rathi N S (2012). Role of women in dairy farming. *Indian Journal of Dairy Science,* **65** (5): 442-446.

Lal R and Khurana A (2011). Gender Issues: The Role of Women in Agriculture Sector. Zenith International Journal of Business Economics & Management Re-search **1**(1):29–39.

Ludgate N (2016). Moser Gender Analysis Framework. Feed The Future.

Lyon S, Bezaury J A and Mutersbaugh T (2010). Gender equity in fair trade-organic coffee producer organizations: Cases from Mesoamerica. Geoforum, **41**(1):93–103. Retrieved from http://farmer.tools4valuechains.org/sites/default/files/documents/Lyon_Aranda_Gender%20and%20Fair%20trade%202010.pdf

Majumder J and Shah P (2017). Mapping the role of women in Indian Agriculture. *Annals of anthropological practice.* **41**(2): 46–54,

March C, Smyth I, and Mukhopadhyay M (1999). A Guide to Gender-Analysis Frameworks, London: Oxfam Publishing

Mencher J P and D'Amico D (1986). Kerala Women as Labourers and Supervi-sors: Implications for Women and Devel-opment. In Visibility and Power: Essays in Women in Society and Development. L., Dube et al., eds. pp. 255–266. New Delhi: Oxford University Press.

Moser C O (1993). Gender Planning and Development: Theory, Practice, and Training. London: Routledge.

Moses V J R E. (2011). "Women empowerment through SHGs : A micro study", International Referred Research Journal, **2**(16).:25.

Murthy R K(Ed). (2001). Building Women's capacities: Interventions in Gender Transformation, New Delhi: Sage Publications.

Myrada Experience, "Self Help Groups, Reading Materials and Related Literature", Gulburga district, 2001, p. 127.

Nag P K, and Nag A. (2004). Drudgery, Accidents and Injuries in Indian Agriculture. Industrial Health. **42**:149–162.

Naila K. (1994). Reversed Realities: Gender Hierarchies in Development Thought. London, UK: Verso.

National Bank for Agriculture and Rural Development (1998). Financing a Self Help Groups. p. 28.

Nayak J, Singh S P and Moharana G. (2013). Occupational Health Hazard Faced by Farm Women at Their Workplaces. pp. 1- 32. Bhubaneswar: DRWA (ICAR).

Netherlands Development Organization, "Gender Reference Guide."

Niamer F M. (1994). Women livestock managers in the third World: focus on technical issues related to gender roles in livestock production. Staff Working Paper 18, IFAD, Rome.

Organization for Security and Co-operation in Europe (2006). Glossary on Gender- Related Terms. Available at: http://www.osce.org/search/apachesolr_search/ glossary%20on%20gender-related%20terms

Orskov E R. (1993). Reality in rural development aid-with emphasis on livestock, Rowett Research Services Ltd., Aberdeen: 1-80.

Overholt C M, Anderson K, Cloud and Austin J. (1985). Gender Roles in Development Projects: Cases for Planners. West Hartford, CT: Kumarian Press.

Parker R. (1993). "Another Point of View: A annual on Gender Analysis Training for Grassroots Workers." New York: UNIFEM.

Potts J, Lynch M, Wilkings A, Huppé G, Cunningham M and Voora V. (2014). The State Of Sustainability Initiatives Review 2014: Standards and the green economy. Winnipeg: IISD. Retrieved from https://www.iisd.org/ pdf/2014/ssi_2014.pdf

Rajendra K K. (1998). "Self help groups and cooperatives Banks", p. 3.

Ratan R P, Topno S K, Topno R K, Singh R K and Bara N. (2005). Effect of Mass Media on Knowledge Gain. Indian Journal of Extension Education. **41**(3&4):80–81.

Rathod P K, Nikam T R, Sariput L, Vajreshwari S and Amit H. (2011). Participation of rural women in dairy farming in Karnataka. *Indian Research Journal of Extension Education*, **11**(2): 31-35.

Reeves D. (2002). Mainstreaming Gender Equality: An Examination of the Gender Sensitivity of Strategic Planning in Great Britain, The Town Planning Review, **73** (2): 197-214. Retrieved from http://agritech.tnau.ac.in/women_in_agri/pdf/index/in_agri.pdf

Retrieved from http://www.energia.org/pubs/papers/g_e_manual/tdg_g_e_mod4.pdf

Retrieved from http://www.ilo.org/public/english/region/asro/mdtmanila/training/unit1/mainmenu.htm

Retrieved from www.censusindia.gov.in

Retrieved from http://agritech.tnau.ac.in/women_in_agri/farm_enter_women_empowerment_durgery_index.htmlhttp://eige.europa.eu/gender-mainstreaming/what-is-gender-mainstreaming

Retrieved from http://drd.nic.in/drd/downloads/programmes-schemes/Gender_Audit.pdf

Retrieved from http://go.illinois.edu/harvardanalyticalframework

Retrieved from http://icds-wcd.nic.in/gbhb/Link%20hand%20pdf/Hand%20Book%20Chap%202.pdf

Retrieved from http://megcnrd.gov.in/forms/SGSY.pdf

Retrieved from http://unesdoc.unesco.org/images/0013/001318/131854e.pdf

Retrieved from www.agricoop.nic.in

Retrieved from www.gdrc.org/gender/framework/framework.html

Retrieved from www.indiastat.com

Retrieved from www.nabard.org

Retrieved from www.ndi.org/files/Guide%20to% 20Gender%20Analysis %20 Frameworks.pdf

Retrieved from www.tnwomen development corporation.org

Retrieved from www.vikaspedia.in

Retrieved from www.vistar.nic.in

Riisgaard L, Michuki G, Gibbon P, Bolwig S, Warring N, and Rantz L. (2009). The performance of voluntary standard schemes from the perspective of small producers in East Africa. Danish Institute for International Studies, on behalf of Traidcraft (UK).

Sah U, Shantanu K and Fulzele R M. (2006). Perceived needs of dairy farmers and farm women related to improved dairy farming in India- an overview. *Agric. Rev.*, **23** (1): 65-70.

Sakal Daily Newspaper (20 June 2008). Emergence of SHG from Amravati

Self employment training center, Amravati.

Sethi R C and Sharma R B. (2011). Frontier Technologies for Empowering Farm Women. International Journal of Farm Sciences **2**(1):142–145.

Sexsmith K. (2017). Promoting Gender Equality in Foreign Agricultural Investments: Lessons from voluntary sustainability standards. Winnipeg: IISD. Retrieved from https://www.iisd.org/sites/default/files/publications/promoting-genderequality-foreign-agricultural-investments.pdf

Sexsmith K, Smaller C, Speller W. (2017). How to Improve Gender Equality in Agriculture. *Investment in agriculture, Policy brief #5*. International Institute for Sustainable Development.

Shailaja R. (2000). Women Energy and Sustainable Develop-ment. Energy for Sustainable Development . **4**(1):45-64.

Sharma K C and Devi L. (2014). Work Time Allocation and Valuation of Women's Contribution in Household Activities in the Urban Economy of Himachal Pradesh. Economic Affairs. **59**(4):687-699.

Sharma J K. (2016). Understanding the Concept of Sensitisation in Humanities and Social Sciences : An Exploration in Philosophy of Mind. International Journal of Scientific Research. **24** (90).: 380-400.

Singh S and Arora R. (2010). Ergonomic Intervention for Preventing Musculoskeletal Disorders among Farm Women. Journal of Agricultural Sciences. **1**(2):61-71.

Singh S. (2007). "Towards inclusives rural financial services - Microfinance in India", Edited by Sangita Kamdar, Microfinance, Self-employment and poverty alleviation, Himalaya Publishing House, p. 61.

Smaller C. (2014). The global response to foreign investment in agriculture (Investment in Agriculture Policy Brief). Winnipeg: IISD. Retrieved from https://www.iisd.org/sites/default/files/publications/global_response_foreign_investment_ag_en.pdf

Smith S. (2013). Assessing the gender impacts of Fairtrade. Social Enterprise Journal, 9(1)., 102-122.

Souza D W. (1999). "Microcredit programme monitoring system : A review of MVRADA experience", Search Bulletin, **14**(13).: 7

Subramanian S. (2010)."A study on self help groups in Tiruneveli district", Manonmaniam Sundavanar University, Tiruneveli.

The World Bank. (2000). Integrating Gender in World Bank Assis-tance. Operations Evaluation Department. The World Bank, Washington, DC.

Thornton P K, Kruska R L, Henninger N, Kristjanson P M, Reid R S and Robinson T P. (2003). Locating poor livestock keepers at the global level for research and development targeting. Land use Policy, **20** (4): 311-322.

Toppo A, Trivedinand M S and Patel A. (2004). Participation of farm women in dairy occupation. *Gujarat Journal of Extension Education,* **15** (2): 15-21.

Utting-Chamorro K. (2007). Does fair trade make a difference? The case of small coffee producers in Nicaragua. Development in Practice, 15(3&4)., 584–599. Retrieved from http://www.tandfonline.com/doi/full/10.1080/09614520500075706?scroll=top&needAccess=true

Verstappen V, Hanson L, and Mclaughlin D. (2012). Gender, health, labor, and inequities: A review of the fair and alternative trade literature. Agriculture and Human Values, 30, 21–39. Retrieved from http://cftn.ca/sites/default/files/AcademicLiterature/Terstappen,%20Hanson%20%26%20Mclaughlin%20online%20first%20ag%20%26%20 Human%20Values%202012.pdf

Waal M. (2006). ¯Evaluating Gender Mainstreaming in Development Projects, Development in Practice, **16**(2).: 209-214.

❑❑❑